George A. Olah

# Carbocations and Electrophilic Reactions

George A. Olah

# Carbocations and Electrophilic Reactions

Verlag Chemie
John Wiley & Sons

Professor Dr. George A. Olah
Department of Chemistry
Case Western Reserve University
Cleveland/Ohio (USA)

First published in Angew. Chem. internat. Edit. *12,* 173 (1973).

With 26 figures and 4 tables

ISBN 3-527-25561-3 (Verlag Chemie, Weinheim)
ISBN 0-471-65314-4 (John Wiley & Sons, New York, NY.)

Library of Congress Catalog Card No. 73-86987.



Printer: Julius Beltz, Hemsbach
Printed in Germany

## Introduction

The general concept of carbocations encompasses all cations of carbon containing compounds which can be differentiated into two distinct, limiting classes: trivalent ("classical") carbenium ions and pentacoordinated ("nonclassical") carbonium ions. In this review the preparation and structural study (by NMR, IR, Raman and ESCA spectroscopy) of stable, long lived carbocations is discussed. Whereas the differentiation of limiting trivalent and pentacoordinated ions serves a useful purpose to establish the significant differences between these ions, it is also clear that in most specific systems there exists a continuum of charge delocalization comprising both intra- and intermolecular interactions.

As is well known, trivalent carbenium ions play an important role in electrophilic reactions of $\pi$- and n-donor systems. Similarly, pentacoordinated carbonium ions are the key to electrophilic reactions of $\sigma$-donor systems (single bonds). The ability of single bonds to act as $\sigma$-donor lies in their ability to form carbonium ions via triangular two electron, threecenter bond formation. Consequently there seems to be in principle no difference between the electrophilic reactions of $\pi$- and $\sigma$-bonds except that the former react more easily, even with weak electrophiles, whereas the latter necessitate more severe conditions. The role of carbocations in electrophilic reactions of $\pi$- and $\sigma$-donor systems is discussed.

The Fieser's cats are by now well-known to most chemists. Maybe for a change some of the dog lovers would like to meet Jim, the Olah's cocker spaniel.

**Contents**

# 1. General Concept

## 1.1. Definition and Differentiation of Carbenium and Carbonium Ions

The concept of carbocations (it seems to be the logical name for all cationic carbon compounds since the negative ions are called carbanions), with exception of the early isolation of highly stabilized triarylcarbenium ion salts, had its origin and grew to maturity through kinetic, stereochemical, and product studies of a wide variety of reactions, especially unimolecular nucleophilic substitutions and eliminations[2]. Leading investigators such as *Meerwein*, *Ingold*, *Hughes*, *Whitmore*, *Bartlett*, *Nenitzescu*, *Winstein*, and others have contributed fundamentally to the development of modern carbenium ion theory, *i. e.* to the concept of electron deficient cationic intermediates. Direct observation of stable, long-lived carbenium ions, generally in highly acidic (superacid) solvent systems, has become possible only in recent years[3].

Based on our continuing study of carbocations by direct observation of long-lived species, it became increasingly apparent that the carbocation concept is much broader than previously realized and necessitates general definition. Such a definition is offered, therefore, based on the realization that *two distinct classes of carbocations exist*[4]:

a) *Trivalent ("classical") carbenium ions* containing an $sp^2$-hybridized *electron deficient* central carbon atom (with six electrons in the valence shell), which, in the absence of constraining skeletal rigidity or steric interference, tends to give a planar (or close to planar) arrangement with the three directly bound atoms. (It should be noted that sp-hybridized, linear acyl cations and vinyl cations also show substantial electron deficiency on carbon.)

b) *Penta-(or tetra)coordinated ("nonclassical") carbonium ions*, which contain five or four coordinated carbon atoms (having eight electrons in the valence shell) bound by three single bonds, and a two-electron, three-center bond (the latter being bound either to two additional bonding atoms or involving a carbon atom to which they are also bound by a single bond).

carbocations

| carbenium ions | carbonium ions |
|---|---|
| trivalent | penta- or tetracoordinated |
| "classical ions" | "nonclassical ions" |
| *e. g.* $CR_3^{\oplus}$ | *e. g.* $CR_5^{\oplus}$ (or $C_2R_5^{\oplus}$) |

Expansion of the carbon octet *via* 3d-orbital participation does not seem possible; there can be only eight valence electrons in the outer shell of carbon[5]. Thus the covalency of carbon cannot exceed four. Pentacoordination implies a species with five or four ligands within reasonable bonding distance from the central atom[6]. The transition states long ago suggested for $S_N2$ and $S_E2$ reactions may represent such cases. However, the direct observation of stable pentacoordinated species in solution was not reported until recently when studies were carried out on long-lived "nonclassical" ions in superacid solvent systems. Moreover, in the past, $S_E2$ substitution reactions have been mainly restricted to organometallic compounds, *e.g.* organomercurials[7].

Interaction of neighboring groups with the vacant p-orbital of the carbenium ion center can contribute to ion stabilization *via* charge delocalization. Such phenomena can involve atoms with unshared electron pairs (n-donors), C—H and C—C hyperconjugation, bent σ-bonds (as in cyclopropylcarbenium ions) and π-electron systems (direct conjugative or allylic stabilization). Thus trivalent carbenium ions can show varying degrees of delocalization without becoming pentacoordinated carbonium ions. The limiting classes defined do not exclude varying degrees of delocalization, but in fact imply a spectrum of carbocation structures.

In contrast to the rather well-defined trivalent ("classical") carbenium ions, "nonclassical ions"[8] have been more loosely defined. In recent years, a lively controversy has centered on the classical-nonclassical carbonium ion problem[9]. The literature contains numerous contributions to the topic, but no generally accepted definition of nonclassical ions has emerged. Furthermore, no direct experimental means have been recognized for the differentiation of classical and nonclassical ions. The extensive use of "dotted lines" when writing carbonium ion structures has been (rightly) criticized by *Brown*[9d], who carried, however, the criticism to questioning the existence of any σ-delocalized (nonclassical) ion. For these ions, if they exist, he stated "...a new bonding concept not yet established in carbon structures is required".

It is suggested that unequivocal experimental evidence has since been obtained for nonclassical ions such as the norbornyl cation[10], and that the bonding concept required to define "nonclassical ions" is to consider them as penta- or tetracoordinated carbonium ions, of which

$CH_5^+$ (the methonium ion, carbonium ion) is the parent, in the same way as $CH_3^+$ (methenium ion, methyl cation, carbenium ion) is the parent for trivalent carbenium ions.

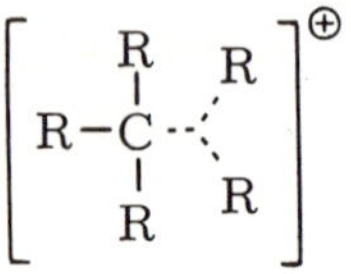

pentacoordinated
R = H or alkyl

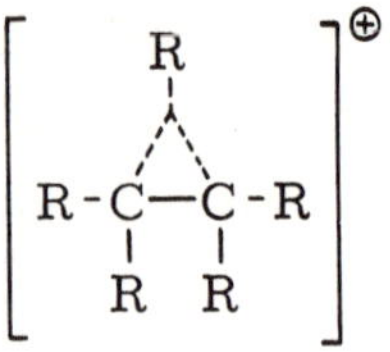

tetracoordinated

For simplicity it is suggested that three-center bonds be depicted by triangular dotted lines drawn from and meeting each other between the three participating atoms; full straight lines are used to symbolize two center, two electron bonds. Such a formulation best illustrates the overlapping bonding orbitals; however, it must be remembered that the point of branching of the dotted line does not represent an additional atom.

Regarding the carbocation concept, it is regrettable that trivalent, planar ions of the type $CH_3^+$ are generally referred to as carbonium ions in the Anglo-Saxon literature. If the name is considered analogous to other onium ions (ammonium, sulfonium, phosphonium ions, *etc.*), then it should relate to the higher valency state carbocation. The higher valency state carbocations, however, clearly are not the trivalent but the *pentacoordinated cations* of the $CH_5^+$ type. In the German and the French literature the term "carbenium ion" is frequently used for naming the trivalent cations. Should we consider these latter ions as protonated carbenes, then this naming is indeed correct[11] and in agreement with IUPAC rules for the nomenclature of organic compounds and we could thus *differentiate* quite simply between *carbenium (trivalent)* and *carbonium (penta- or tetracoordinated) ions*[12]. It should be pointed out, however, that "carbenium ion" should be used only for the trivalent ions and not as a generic name for all carbocations. Owing to the wide acceptance of naming all cations of carbon, as carbonium or carbenium ions, we find the use of prefixes for differentiating trivalent and penta- or tetracoordinated is advisable. However, in the future the nomenclature either must differentiate between trivalent carbenium ions and pentacoordinated carbonium ions, or use the carbo*cation* naming system.

## 1.2. Development of Modern Theory: Kinetic and Stereochemical Studies

One of the most daring and fruitful ideas born in organic chemistry was the suggestion that carbocations might be intermediates in reactions that start from nonionic reactants and lead to nonionic covalent products. It was *Meerwein*[13] who in 1922, while studying the kinetics of the rearrangement of camphene hydrochloride to isobornyl chloride, reported the important observation that the reaction rate increased in a general way with the dielectric constant of the solvent. Further, he found that metal chlorides—such as $SbCl_5$, $SnCl_4$, $FeCl_3$, $AlCl_3$, and $SbCl_3$ (but not $BCl_3$ or $SiCl_4$), as well as dry HCl—which promote the ionization of triphenylmethyl chloride by formation of ionized complexes, considerably accelerate the rearrangement of camphene hydrochloride. *Meerwein* concluded that the conversion of camphene hydrochloride into isobornyl chloride actually does not proceed by way of migration of the chlorine atom but by a rearrangement of a cationic intermediate. Thus the modern concept of carbocations was born.

In England, starting in the late 1920's, *Ingold* and *Hughes et al.* carried out detailed kinetic investigations on what later became known as nucleophilic substitution at saturated carbon and on polar elimination reactions[14]. The well-known work relating to $S_N1$ and later to E1 reactions established the carbocation concept. In a series of papers from 1932 onwards *Whitmore*[15] extended *Meerwein*'s rearrangement theory to many organic chemical reactions.

Kinetic and stereochemical evidence helped to establish the role played by carbocations in organic reactions. Generally, however, they were considered only as very short lived species and could not be directly detected by physical means.

### 1.3. Observation of Stable, Long-lived Ions

The transient nature of carbocations in chemical reactions arises from their extreme reactivity with nucleophiles. By the use of highly acidic and very weakly nucleophilic solvents, it is possible to prevent reactions leading to covalent products; in the last decade or so it has been possible in this way to observe a wide range of *stable*, long-lived carbocations.

# 2. Preparation and Structure of Stable Trivalent Carbenium Ions

## 2.1. Alkylcarbenium Ions (Alkyl Cations)

### 2.1.1. Early Unsuccessful experiments

Simple alkylcarbenium ions were, until recently, considered only as transient entities[2] and their existence has been inferred from studies on the course of certain reactions. No reliable physical or chemical methods other than that of electron impact measurements were known for the detection of simple alkylcarbenium ions. The formation of organic cations in the gaseous phase by electron bombardment of alkanes, haloalkanes, and other precursors has been widely investigated by mass spectrometry[16]. An analogous direct observation of carbenium ions in solution had not been achieved.

Observation of alkylcarbenium ions (trivalent alkyl cations) such as the trimethylcarbenium ion (*tert*-butyl cation) $(CH_3)_3C^+$ or the dimethylcarbenium ion (isopropyl cation) $(CH_3)_2CH^+$ has thus been a long standing challenge for many research workers. The occurrence of alkylcarbenium ions in systems of alkyl halides and Lewis acid halides has been inferred from a variety of observations, such as those of vapor pressure depressions of $CH_3Cl$ and $C_2H_5Cl$ in the presence of gallium chloride[17], the electrical conductivity of aluminum chloride in alkyl chlorides[18] and of alkyl fluorides in boron trifluoride[19], as well as the effect of ethyl bromide on the dipole moment of aluminum bromide[20]. However, in no case had a well-defined, stable ionic salt or complex been established, even at very low temperatures.

*Rosenbaum* and *Symons*[21] recorded the electronic spectra of alcohols and olefins in sulfuric acid. They observed, for a number of simple aliphatic alcohols and olefins, absorption maxima around 290 nm which they ascribed to the corresponding alkylcarbenium ions.

Later, *Finch* and *Symons*[22] reinvestigated the absorption of aliphatic alcohols and olefins in sulfuric acid, and showed that the condensation products formed with acetic acid (used as solvent for the precursor alcohols and olefins) were responsible for the spectra, not the simple alkylcarbenium ions. Protonated mesityl oxide was identified as the species responsible for the absorption at 290 nm in the isobutylene/acetic acid/sulfuric acid system.

*Deno et al.*[23] carried out an extensive study of the fate of alkyl cations in $H_2SO_4$ and oleum. They found that equal amounts of a mixture of saturated hydrocarbons ($C_4$ to $C_{18}$) that are insoluble in $H_2SO_4$ and a mixture of cyclopentenyl cations ($C_9$ to $C_{20}$) that are soluble in the $H_2SO_4$ layer are formed. Cyclopentenyl cations exhibit strong ultraviolet absorption at about 300 nm.

It must be concluded, therefore, that earlier attempts to prove the existence of stable, well-defined alkylcarbenium ions were unsuccessful, both in experiments with sulfuric acid solutions as well as with systems from alkyl halides and Lewis acid halides.

### 2.1.2. Preparation of Carbenium Ions from Alkyl Fluorides in Solutions Containing Antimony Pentafluoride

In 1962 we reported the first direct observation of stable alkylcarbenium ions in solution[24–26]. We obtained the trimethylcarbenium ion (*tert*-butyl cation) when *tert*-butyl fluoride was dissolved in excess antimony pentafluoride, which served both as Lewis acid and solvent. Later the counter ion was found to be, under these conditions, primarily the dimeric anion $Sb_2F_{11}^-$[27]; whereas in $SbF_5$—$SO_2$ or $SbF_5$—$SO_2ClF$ solutions, both $SbF_6^-$ and $Sb_2F_{11}^-$ were found.

$$(CH_3)_3CF + (SbF_5)_2 \rightarrow (CH_3)_3C^{\oplus}Sb_2F_{11}^{\ominus}$$

The possibility of obtaining stable alkylcarbenium fluoroantimonate salts from alkyl fluorides (and subsequently other halides) in antimony pentafluoride solution (neat or diluted with sulfur dioxide, sulfuryl chloride fluoride, or sulfuryl fluoride) or in other superacids[28] [particularly $FSO_3H$—$SbF_5$ (magic acid®[29]), HF—$SbF_5$, fluoroantimonic acid, see subsequent discussion) was investigated in detail, our studies extending to all isomeric $C_3$, $C_4$ and $C_5$ alkyl fluorides, as well to a number of higher alkyl halides.

Propyl, butyl, and pentyl fluorides in excess antimony pentafluoride gave the isopropyl *(1)*, *tert*-butyl *(2)*, and *tert*-amyl cations *(3)* (as their fluoroantimonates).

$$C_3H_7F + (SbF_5)_2 \longrightarrow H_3C\text{-}\overset{\oplus}{C}H\text{-}CH_3 \quad Sb_2F_{11}^{\ominus}$$

*(1)*

$$C_4H_9F + (SbF_5)_2 \longrightarrow H_3C\text{-}\overset{\oplus}{C}(CH_3)\text{-}CH_3 \quad Sb_2F_{11}^{\ominus}$$

*(2)*

$$C_5H_{11}F + (SbF_5)_2 \longrightarrow H_3C\text{-}\overset{\oplus}{C}(CH_3)\text{-}C_2H_5 \quad Sb_2F_{11}^{\ominus}$$

*(3)*

The secondary butyl and amyl cations can be observed at low temperatures, but they rearrange readily to more stable tertiary ions.

Generally, the most stable tertiary or secondary carbenium ions are eventually observed on starting from any of the isomeric alkyl fluorides in superacidic solvent systems. Rearrangement to these more stable ions can be studied, starting from very low temperatures.

### 2.1.3. NMR-Spectra

One of the most powerful tools in the study of carbenium ions is nuclear magnetic resonance spectroscopy. The main feature of the proton NMR spectra of alkyl fluorides in antimony pentafluoride is the substantial deshielding of the

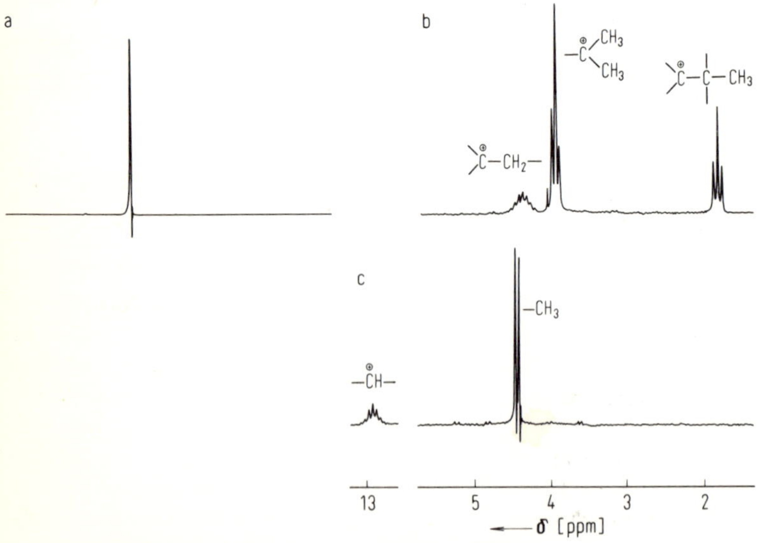


Fig. 1. ¹H-NMR spectra of: a) the *tert*-butyl cation *(2)* [trimethylcarbenium ion, $(CH_3)_3C^+$]; b) the *tert*-amyl cation *(3)* [dimethylethylcarbenium ion, $(CH_3)_2C^+—C_2H_5$]; c) the isopropyl cation *(1)* [dimethylcarbenium ion, $(CH_3)_2C^+H$]. (60 MHz, $SbF_5—SO_2ClF$ solution, −60°C.)

protons in the carbenium ions as compared with in the starting alkyl fluorides (Fig. 1 and Table 1) and the absence of H—F coupling.

In order to prove that stable alkylcarbenium ions, and not exchanging donor-acceptor complexes were obtained, we investigated the $^{13}C$ nuclear magnetic resonance of the potentially electropositive carbenium carbon atom in alkylcarbenium ions[30].

The $^{13}C$ chemical shift of $C^+$ in the *tert*-butyl cation $(CH_3)_3{}^{13}C^+$ in $SO_2ClF—SbF_5$ solution at $-20\,°C$ is $\delta_{^{13}C}$ $-135.4$ (with reference to capillary $^{13}CS_2$, as given

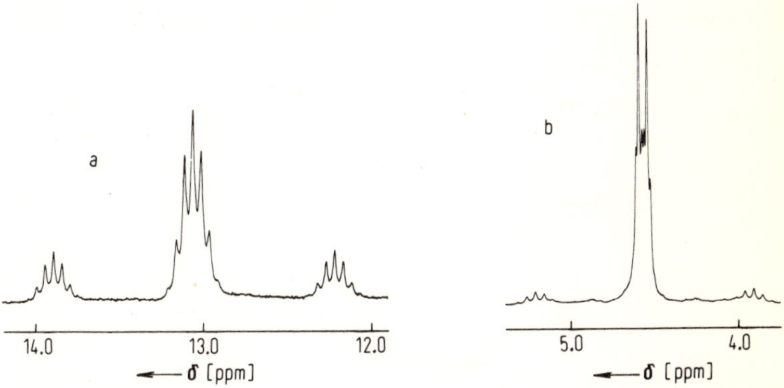


Fig. 2. 100-MHz $^1H$-NMR spectrum of the isopropyl cation generated from isopropyl chloride with a 50% $^{13}C$ enrichment of C-2, in $SO_2ClF—SbF_5$ solution. The spectrum was recorded at $-60\,°C$. Figure 2a shows the methine region and consists of a septuplet ($J_{HH} = 5$ Hz) with $^{13}C$ satellites 84.5 Hz on either side of the main resonance ($J_{CH} = 169$ Hz). Figure 2b shows the methyl region. The $^{12}C$ species gives a doublet in this region while the $^{13}C$-labeled ions give a doublet of doublets due to the long-range CCH coupling of 3.3 Hz. Some scrambling of the $^{13}C$ label into the methyl groups is apparent in this spectrum giving small pentuplets 65.5 Hz on either side of the main resonance. The structure of these satellites arises from the fact that the three- and four-bond proton-proton coupling constants are similar in magnitude (5 and 6 Hz).

throughout this review), the long-range coupling constant ($J_{CCH}$) to the methyl protons being 3.6 Hz.

Under identical conditions the $^{13}C$ chemical shift of the $C^+$ in the isopropyl cation is $\delta_{^{13}C}$ −125.0, and the long-range coupling to the methyl protons is 3.3 Hz (Fig. 2).

The direct $^{13}C$—H coupling is 169 Hz (a good indication of the $sp^2$ hybridization of the carbenium carbon atom is obtained from the apparent linear relation of the coupling constant with the hybridization; the electron-deficient carbon atom shows 33.8% s-character). The long-range, proton-proton coupling constant is 6.0 Hz (see Fig. 3).

Substitution of the methyl group in the *tert*-butyl cation by hydrogen thus causes a shielding of the $C^+$ (upfield shift of the $^{13}C^+$ signal of about 10.4 ppm). If a direct relation between $^{13}C$ shifts and electron density at $C^+$

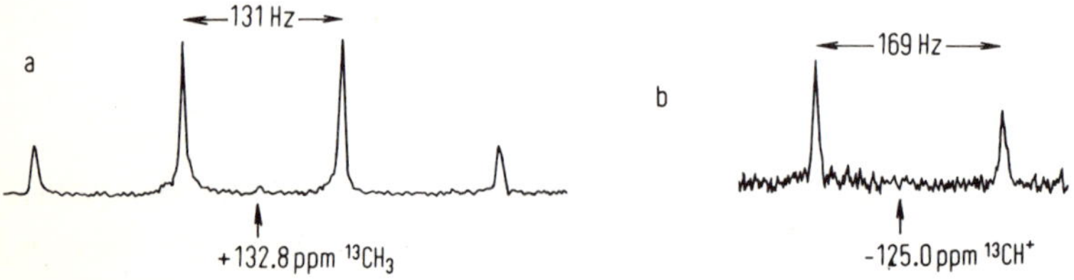


Fig. 3. 25-MHz $^{13}C$-INDOR spectrum [30a] of the isopropyl cation. The spectra were obtained at −20 °C by monitoring the signal intensity of the $^{13}C$ satellites shown in Figure 2 while sweeping the irradiating frequency. The peaks observed are the result of splitting of the satellite due to "tickling" of connected $^{13}C$ transitions. Figure 3a shows the quartet corresponding to the methyl $^{13}C$ and Figure 3b the doublet from the methine $^{13}C$. The latter spectrum was obtained from the low-field $^{13}C$ satellite in the proton spectrum. The low-field $^{13}C$ transition is connected regressively and the high-field $^{13}C$ transition connected progressively to this transition leading to the observed asymmetry of the $^{13}C$-INDOR spectrum. This asymmetry is particularly apparent in C-INDOR spectra obtained using a low-amplitude "tickling" field.

exists, then it could be taken that the central carbon atom in the *tert*-butyl cation is slightly more positive than that in the isopropyl cation. Otherwise the difference can be accounted for in terms of a substituent effect (*i.e.* $CH_3$ *vs.* H).

The $^{13}C$ chemical shift of $C^+$ in the *tert*-amyl cation $[C_2H_5C^+(CH_3)_2]$ is $\delta_{^{13}C}$ $-139.4$, *i.e.* $C^+$ is more strongly deshielded (shift downfield of *ca.* 4 ppm) than $C^+$ in the *tert*-butyl cation. This shift difference is much smaller than the 17 ppm found in the case of the related alkanes, although the shift observed is in the same direction.

It should, however, be pointed out that other factors than electron density can also affect the chemical shifts; thus no quantitative conclusion relating to charge densities can be established from the NMR data.

The total shielding constant $\sigma$ for a particular nucleus can be regarded as the sum of three terms

$$\sigma = \sigma_d + \sigma_p + \sigma'$$

where $\sigma_d$ is the diamagnetic term, $\sigma_p$ the paramagnetic term, and $\sigma'$ the term arising from neighboring groups. For a carbon nucleus the dominant paramagnetic contribution is affected principally by charge polarization, variation in bond order, and the average excitation energy. The relative significance of each factor must be considered.

It is interesting to note that theoretical calculations indicate higher positive charge density on the carbenium center in the *tert*-butyl cation, than on the carbenium center of the isopropyl cation (naturally, this relates only to the gas phase).

It is difficult to interpret the observed large deshieldings and coupling-constant values in any way other than as a direct proof that (i) the central carbon atoms involved in the carbenium ion are $sp^2$-hybridized, and (ii) that, at the same time, the carbon atom carries a substantial positive charge.

Table 1 lists the parameters of the $^1$H- and $^{13}$C-NMR spectra of the cationic centers of a few secondary and tertiary carbenium ions. The main points to be noted are the downfield chemical shifts ($\delta_{^{13}C}$) which point to extensive positive character and the coupling constants ($J_{CH}$) indicative of $sp^2$-hybridization.

When the isopropyl cation was generated from 50%-labeled [2-$^{13}$C]-2-chloropropane by dissolution in $SO_2ClF$—$SbF_5$ at −60 °C, equilibration of the $^{13}$C labeling occurred with a half-life of 1 hour; after several hours the $^{13}$C labeling was distributed equally among the three carbons ("scrambling"). This observation suggests involve-

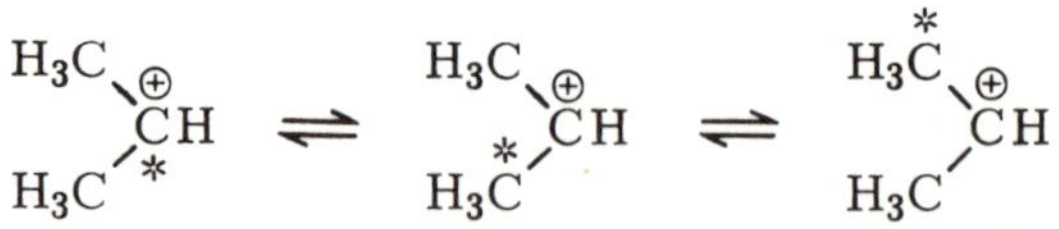


ment of protonated cyclopropanes in the carbon scrambling process (see below). Similar scrambling was observed in the *sec*-butyl and *tert*-amyl cations (*Saunders*[31]).

## 2.1.4. Infrared and Raman Spectra

Infrared and Raman spectra of the stable alkylcarbenium ions were also recorded[25, 32] and are in complete agreement with the carbenium structure of the ions. The IR

Table 1. Characteristic NMR parameters of alkylcarbenium ions in $SbF_5$—$SO_2ClF$ solution at −70 °C.

| Ion | $^{1}$H-NMR (δ [ppm]) CH$^{\oplus}$ | $J_{\oplus CH}$ | $J_{\oplus CCH}$ | α-$CH_2$ | α-$CH_3$ | β-$CH_3$ | $^{13}$C-NMR (δ [ppm]) C$^{\oplus}$ | α-$CH_3$ |
|---|---|---|---|---|---|---|---|---|
| $(CH_3)_2CH^{\oplus}$ *(1)* | 13 | 169 | 3.3 | | 4.5 | | −125.0 | 132.8 |
| $(CH_3)_3C^{\oplus}$ *(2)* | | | 3.6 | | 4.15 | | −135.4 | 146.3 |
| $(CH_3)_2C^{\oplus}C_2H_5$ *(3)* | | | | 4.5 | 4.1 | 1.94 | −139.2 | 150.1 |
| $H_3CC^{\oplus}(C_2H_5)_2$ *(4)* | | | | 4.44 | 4.16 | 1.87 | −139.4 | 150.8 |

Table 2. Raman and IR frequencies of the *tert*-butyl cation and [D$_9$]-*tert*-butyl cation and their correlation with those of $(CH_3)_3B$ and $(CD_3)_3B$.

| Species | Frequency of vibration [cm$^{-1}$] $\nu_1, \nu_{12}, \nu_7, \nu_{19}$ | $\nu_2, \nu_{13}$ | $\nu_{21}$ | $\nu_{14}$ | $\nu_{15}$ | $\nu_{17}$ | $\nu_5$ | $\nu_{16}$ | $\nu_6$ | $\nu_9$ | $\nu_{10}$ | $\nu_{18}$ |
|---|---|---|---|---|---|---|---|---|---|---|---|---|
| $(CH_3)_3C^{\oplus}$ *(2)* | 2947 | 2850 | | 1450 | | 1295 | | | 667 | | 347 | 306 |
| $(CH_3)_3B$ | 2975 | 2875 | 1060 | 1440 | 1300 | 1150 | 906 | 866 | 675 | 973 (486?) | 336 [a] | 320 |
| $(CD_3)C^{\oplus}$ | 2187 | 2090 | | 1075 | | 980 | | | 720 | | 347 | 300 |
| $(CD)_3B$ | 2230 | 2185 | | 1033 | 1018 | 1205 | | | 620 | 870 | (289) [b] | (276) [b] |

[a] IR frequency.
[b] Calculated.

spectra of these ions and of their deuterated analogs correspond to the spectra predicted by calculations based on selected molecular models and force constants. Thus, vibrational spectra can also be used in the identification of stable carbenium ions.

Laser Raman spectroscopy is another powerful tool in the study of carbocations. Since Raman spectra give valu-

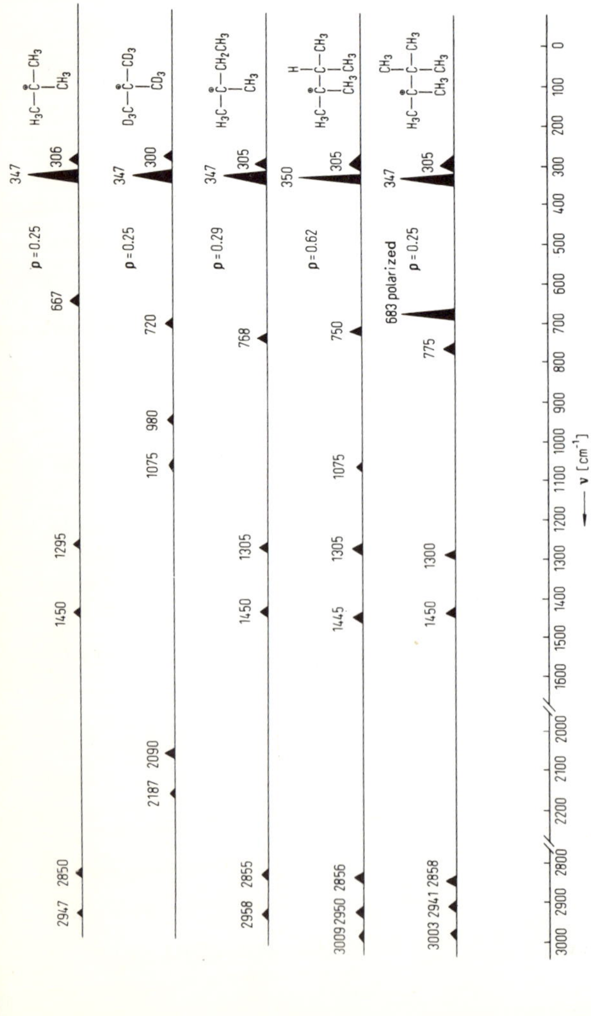

Fig. 4. Schematic representation of Raman spectra of alkyl cations

able information on symmetry, these spectra help to establish in detail the structures of the ions and their configurations. Raman spectroscopic data provide strong evidence, for example, that the trimethylcarbenium ion in magic acid[29] prefers a conformation leading to overall $C_{3v}$ point group symmetry (Table 2). Thus, the $\overset{+}{C}(CH_3)_3$ ion possesses a planar $C^+(C_3)$ carbon skeleton with one hydrogen atom of each $CH_3$ group above the plane of the carbon atoms. The other two hydrogen atoms are arranged symmetrically below this plane to the right and left of the $C_3$ axis. Raman spectra observed for the *tert*-amyl cation (dimethylethylcarbenium ion), the pentamethylethyl cation (dimethyl-*tert*-butylcarbenium ion), and the dimethylisopropylcarbenium ion also show similar structure (Fig. 4). Raman and IR spectroscopic studies thus provide, in addition to $^{13}$C-NMR data, direct evidence for the planar (or near planar) carbenium center of alkylcarbenium ions.

Evidence for planarity or near planarity of the $sp^2$ center of trivalent alkylcarbenium ions thus comes from the combined results of NMR ($^1$H and $^{13}$C), IR, and Raman spectroscopy[30,32].

### 2.1.5. UV Spectra

The observation of stable alkylcarbenium ions in antimony pentafluoride and related superacid solutions also opened up the possibility of investigating the UV and visible spectra of such solutions. *Waack*, *Pittman*, and *I* reported[33] that solutions of alkylcarbenium ions in $FSO_3H$—$SbF_5$ solutions at $-60\,°C$ showed no absorption maxima above 210 nm. In view of this observation, it is highly probable that previous claims relating to a 290 nm absorption of alcohols and olefins in sulfuric acid solutions were due to condensation products or cyclic allylic ions and not to the simple alkylcarbenium ions[34].

### 2.1.6. ESCA Spectra

ESCA spectroscopy[35] is an extremely useful method for the investigation of carbocations[36]. Within such molecules the positive charge is generally unequally distributed over different atoms of the carbon skeleton. Consequently, the core electrons of these atoms are differently screened and show increasing binding energies with increasing positive charge on the C atoms. With correspondingly large positive charge on the carbocationic center the energy differences are large enough to give rise to separate 1s-level photoelectron lines.

In contrast measurements of carbon 1s electron binding energies in several hydrocarbons yielded a rather narrow range of chemical shifts. For instance, one cannot distinguish between the carbon atoms of neopentane and those of benzene, for their 1s binding energies are virtually the same (290.4 eV). Ethane (290.6 eV), ethylene (290.7 eV), and acetylene (291.2 eV) carbon 1s binding energy differences are well below the presently attainable resolution with ESCA measurements.

These results are in agreement with theoretical expectations. Core electrons are barely affected unless their screening against nuclear attraction is modified by a significant change in the outer (valence) electron shell. In other words, core electron binding energies are mainly dependent on the formal charge of the corresponding atom and on the electronegativity of attached atoms or groups of atoms.

The fact that such factors are minimal in hydrocarbons accounts for the small differences described above.

Since ESCA spectroscopy is a 'surface' method the recording of the spectra of the carbenium ions was carried out on frozen superacid solutions or on isolated salts; they were recorded at liquid nitrogen temperature[36].

The binding energies $E_b$ (defined as differences between the Fermi level and the 1s atomic level energies) are given by the equation

$$E_b = E_{h\nu} = E_{kin} - \Phi_s$$

where $E_{h\nu}$ is the energy of the exciting radiation (*e.g.* $Al_{K\alpha}$ X-rays), $E_{kin}$ is the measured kinetic energy of the photoejected electron, and $\Phi_s$ is the spectrometer work function (the energy necessary to bring the electron from the Fermi level to the free-electron level). The 1s-electron spectrum (ESCA spectrum) of the *tert*-butyl cation is shown in Figure 5.

The upper trace in Figure 5 exhibits two clearly separated peaks with a separation of $dE_{b+C-C} = 4 \pm 0.2$ eV. The lower trace represents the result given by a curve resolver. The peak area ratio is 1 : 3.

The experimentally determined carbon 1s binding energy difference between the carbenium ion center and the remaining three carbon atoms lies within the limit of that predicted by *ab initio* quantum mechanical calculations (4.45 eV). Comparable results were obtained in experiments carried out in collaboration with *Mateescu* on a series of other alkylcarbenium ions, such as the *tert*-amyl cation *(3)* ($dE_{b+C-C} = 4 \pm 0.2$ eV).

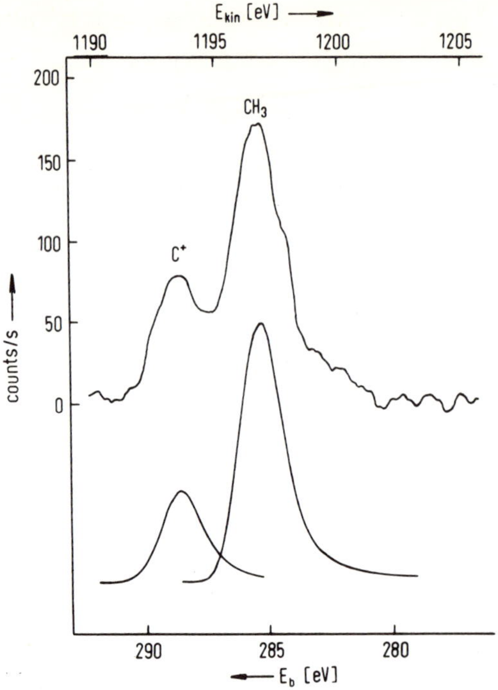


Fig. 5. Carbon 1s electron spectrum (ESCA) of the *tert*-butyl cation *(2)*. $E_b$ = binding energy, $E_{kin}$ = kinetic energy.

## 2.1.7. Preparation of Carbenium Ions from Other Precursors

Alkylcarbenium ions can be formed not only from halides (the earlier investigation of generation from alkyl fluorides was later extended to alkyl chlorides, bromides, and even iodides) but also from other precursors such as olefins in superacids like HF—$SbF_5$ and $FSO_3H$—$SbF_5$[37, 38].

$$RCH{=}CH_2 \xrightarrow[HF\text{-}SbF_5]{FSO_3H\text{-}SbF_5} R\overset{\oplus}{C}H\text{-}CH_3$$

Tertiary and reactive secondary alcohols in superacids like $FSO_3H—SbF_5$ (magic acid®[29])[34], $FSO_3H$, and $SbF_5—SO_2—(SO_2ClF)$ also ionize to the corresponding stable carbenium ions[39]. The generation of alkylcarbenium ions from alcohols indicates the great advantages of increasing acidity and of using acid systems with low freezing points. *Deno*[23] showed that the use of sulfuric acid and oleum results in formation of cyclized allylic ions from simple aliphatic alcohols. If the extremely strong acid $FSO_3H—SbF_5$ ("magic acid"®) is used these alcohols can be ionized to the corresponding tertiary and secondary alkylcarbenium ions without further rearrangements taking place.

$$(CH_3)_3C\text{-}OH \xrightarrow[-H_2O]{FSO_3H\text{-}SbF_5} (CH_3)_3C^{\oplus}$$

Primary and less reactive secondary alcohols are protonated in $FSO_3H—SbF_5$ solution at low temperatures (−60 °C) and show very slow or no proton exchange with the superacid[40].

$$H_3C\text{-}CH_2\text{-}CH_2OH \underset{-H^{\oplus}}{\overset{+H^{\oplus}}{\rightleftharpoons}} H_3C\text{-}CH_2\text{-}CH_2\overset{\oplus}{O}H_2$$

$$H_3C\text{-}\underset{\substack{|\\ OH}}{C}H\text{-}CH_3 \underset{-H^{\oplus}}{\overset{+H^{\oplus}}{\rightleftharpoons}} H_3C\text{-}\underset{\substack{|\\ {}^{\oplus}OH_2}}{C}H\text{-}CH_3$$

Temperature-dependence studies of the NMR spectra allow the kinetics of dehydration of protonated alcohols to the corresponding carbenium ions to be followed[41].

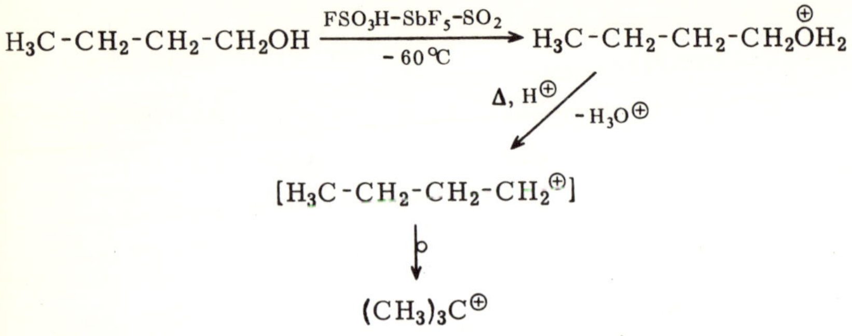

Antimony pentafluoride itself (neat or in $SO_2$ or $SO_2ClF$ solution) ionizes alcohols smoothly to alkylcarbenium ions.

$$R{-}OH \xrightarrow{SbF_5} R^{\oplus}SbF_5OH^{\ominus}$$

We have found that difficulties in the ionization of primary (and less reactive secondary) alcohols at low temperatures can be overcome, in some cases, if the alcohols are first converted with thionyl halides or carbonyl halides into the corresponding haloformates *(5)* or halosulfites *(6)*, which then readily ionize in $SbF_5{-}SO_2$ solution and lose $CO_2$ or $SO_2$ (fragmentative ionization)[37,42].

$$ROH \xrightarrow{COX_2} RO{-}\overset{\overset{O}{\|}}{C}{-}X \;(5) \xrightarrow{SbF_5{-}SO_2} R^{\oplus}SbF_5X^{\ominus} + CO_2$$

$$ROH \xrightarrow{SOX_2} RO{-}\overset{\overset{O}{\|}}{S}{-}X \;(6) \xrightarrow{SbF_5{-}SO_2} R^{\oplus}SbF_5X^{\ominus} + SO_2$$

$$X = Cl, F$$

Aliphatic ethers such as *(7)* are protonated in strong acids, and subsequently cleaved (at higher temperatures) to alkylcarbenium ions[43].

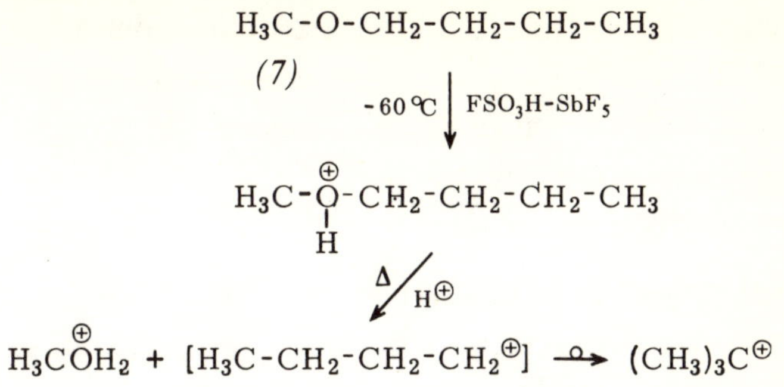


The protonation and subsequent ionization of mercaptans (thiols) and sulfides have been studied in a similar way[44].

Superacids such as $FSO_3H—SbF_5$ act as very effective hydrogen abstracting agents, allowing the generation of carbocations from saturated hydrocarbons[45].

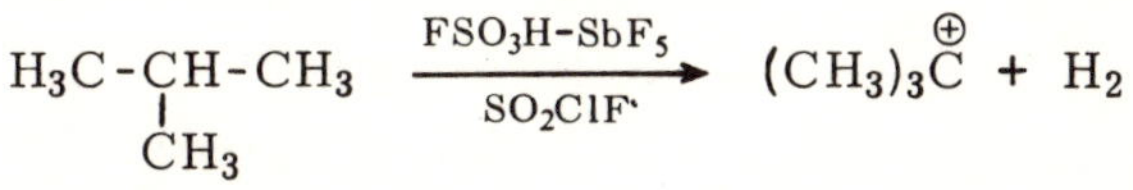


Antimony pentafluoride itself (without addition of $FSO_3H$ or HF) also ionizes alkanes to the corresponding alkylcarbenium ions. Since $SbF_5$ always contains traces of HF, however, it is questionable whether the reaction is effected by the Lewis acid itself.

Carbenium ions can also be generated by decarbonylation of *tert*-alkyloxocarbenium ions, such as the *tert*-butyloxocarbenium ion (pivaloyl cation) *(8)*[26].

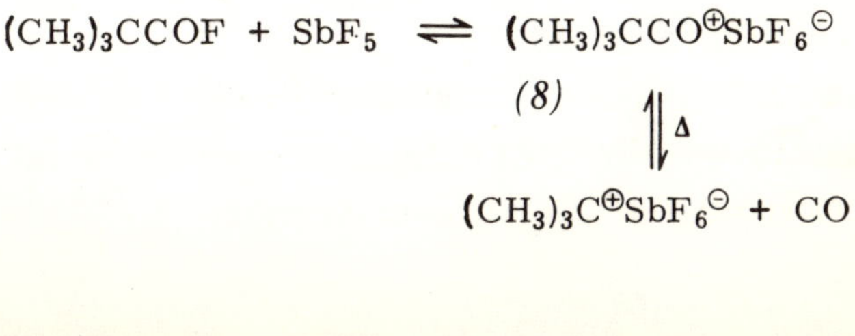

This reaction corresponds to the reverse of the Koch-Haaf acid synthesis, which is known to involve reaction of carbenium ions with carbon monoxide. Indeed, the pivaloyl cation can be obtained by reaction of the tertiary butyl cation with carbon monoxide[26,46].

Thiols and thioesters (sulfides), like their oxygen analogs, can also be used as starting substances for preparation of alkylcarbenium ions[47]; their ionization in superacids generally necessitates more forcing conditions (higher temperatures). *S*-Alkyl halothioformates *(9)* also form carbenium ions *via* fragmentative ionization[48].

$$\mathrm{RSH} \xrightarrow[\mathrm{HF{-}SbF_5,\ FSO_3H{-}SbF_5}]{\mathrm{SbF_5\ or}} \mathrm{R^{\oplus}SbF_5SH^{\ominus}}$$

$$\underset{(9)}{\mathrm{RS{-}C({=}O){-}Cl(F)}} \xrightarrow{\mathrm{SbF_5{-}SO_2ClF}} \mathrm{R^{\oplus}SbF_5Cl^{\ominus}(F)} + \mathrm{COS}$$

Amines also can be used as precursors for the generation of alkylcarbenium ions. The classical method of deaminative formation of carbenium ions involves some type of diazotization reaction producing an equimolar amount of water.

$$\mathrm{RNH_2} \xrightarrow[-\mathrm{H_2O}]{\mathrm{HNO_2}} [\mathrm{RN_2^{\oplus}}] \longrightarrow \mathrm{R^{\oplus}} + \mathrm{N_2}$$

This difficulty can be overcome by first converting the amine into the corresponding sulfinylamine *(10)* or isocyanate *(11)*, which is then reacted with stable nitrosyl ($NO^+$) salts to give the related carbenium ion with evolution of $N_2$ and $SO_2$ (or $CO_2$) instead of water[49].

$$RNH_2 \xrightarrow{SOCl_2} \underset{(10)}{RNSO} + NO^{\oplus}SbF_6^{\ominus} \rightarrow R^{\oplus}SbF_6^{\ominus} + N_2 + SO_2$$

$$RNH_2 \xrightarrow{COCl_2} \underset{(11)}{RNCO} + NO^{\oplus}SbF_6^{\ominus} \rightarrow R^{\oplus}SbF_6^{\ominus} + N_2 + CO_2$$

### 2.1.8. Preparation of Alkylcarbenium Ions in Superacids Not Containing Antimony Pentafluoride

Whereas superacids (such as $HF—SbF_5$, $FSO_3H—SbF_5$, $CF_3SO_3H—SbF_5$ *etc.*) containing antimony pentafluoride are the preferred solvents for obtaining stable alkylcarbenium ions, other superacids such as $HF—BF_3$, $HF—TaF_5$ or fluorosulfuric acid can, on occasions, also be used successfully. The stability of carbocations—with exception of tertiary ions—in these solvents is, however, generally lower.

## 2.2. Cyclopropylcarbenium Ions

Solvolysis studies carried out by *Roberts*[50] and *Hart*[51] showed both the unusual stability of cyclopropylcarbenium ions and the facility with which such ions rearrange.

Cyclopropyl groups bound to the carbocation center have a strong stabilizing effect and effectively delocalize charge. The direct observation of cyclopropylcarbenium ions by NMR spectroscopy provides clear indication of charge delocalization.

The first cyclopropylcarbenium ion observed directly was the tricyclopropylcarbenium ion *(12)*[52]. Its NMR spectrum in $H_2SO_4$ consists of a single sharp line at $\delta$ 2.26.

In the 300 MHz spectrum of a solution in $SbF_5$—$SO_2ClF$, however, the signals of the methine and methylene cyclopropyl protons are well resolved[53] (Fig. 6a).

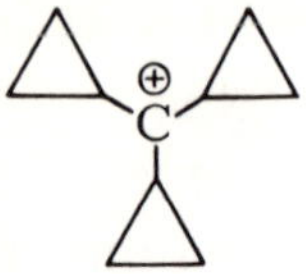

*(12)*

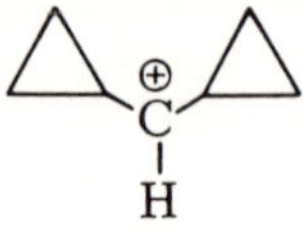

*(13)*

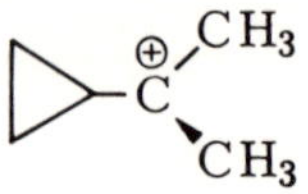

*(14)*

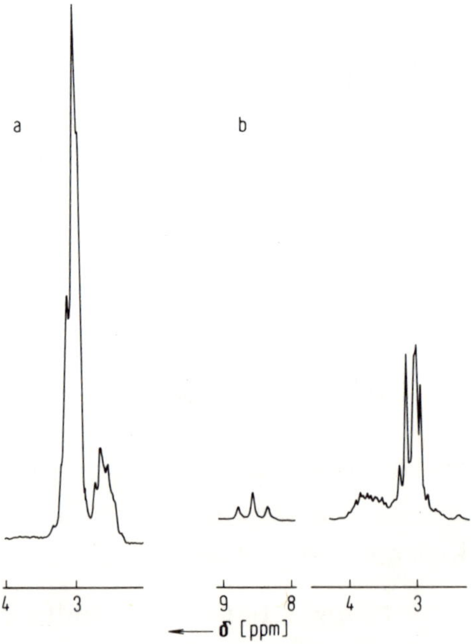


Fig. 6. a) 300-MHz $^1$H-NMR spectrum of the tricyclopropylcarbenium ion *(12)*; b) 60-MHz $^1$H-NMR spectrum of the dicyclopropylcarbenium ion *(13)*.

*Pittman*[54] and *I* have observed several tertiary and secondary cyclopropylcarbenium ions, such as the dicyclopropylcarbenium ion *(13)* (Fig. 6b). The solvent used was usually

$SbF_5$—$SO_2$ or $FSO_3H$—$SbF_5$—$SO_2$. The ring protons are resolved in the NMR spectrum of each of these ions. Most interesting of these ions is the dimethylcyclopropylcarbenium ion *(14)* (Fig. 7). The two methyl groups are non-equivalent and are separated by 0.54 ppm. Since in uncharged systems the hydrogens lying in the face of the cyclopropane ring experience an upfield shift of 0.3 to 0.5 ppm, we concluded that the cyclopropane ring lies in a plane perpendicular to the plane of the dimethylcarbenium ion center (bisected form).

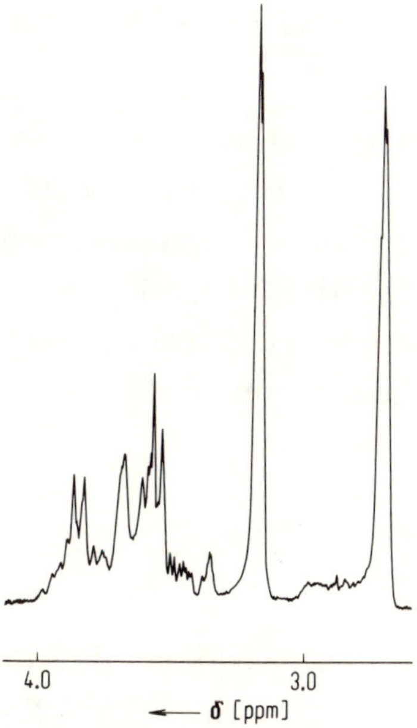


Fig. 7. 100-MHz $^1$H-NMR spectrum of the dimethylcyclopropylcarbenium ion *(14)*.

The plane of the cyclopropane ring lies parallel to the axis of the vacant p orbital of the carbenium carbon atom. In this configuration, one methyl group is *cis* to the cyclopropane ring and the other is *trans*. The *cis* methyl group now experiences the diamagnetic anisotropy of the cyclopropane ring; this accounts for its observed position 0.54 ppm upfield from the *trans* methyl group. The methyl bands do not coalesce at −30 °C, at which temperature ring opening takes place in $FSO_3H$—$SbF_5$ solution and the ion undergoes subsequent decomposition reactions. This shows that the cyclopropane ring does not rotate at this temperature, and that the rotational barrier is at least 10 kilocalories per mol. Recently, *Kabakoff* and *Namanworth*[54a] succeeded, using double resonance experiments, in determining the barrier as 13.7 kcal/mol (in $SbF_5$—$SO_2ClF$).

In contrast to the "classical" tertiary and secondary cyclopropylcarbenium ions *(12)*—*(14)* (showing of course substantial charge delocalization over the cyclopropane ring but maintaining the identity of the trivalent ions), primary cyclopropylmethyl cations show entirely different non-classical carbonium ion character (see Section 3).

### 2.3. Alkenyl Cations

A large number of alkenyl cations has been directly observed, particularly by *Deno* and *Richey*[23, 55], *Sorensen*[56], and *Olah*[57] and their co-workers. *Deno* reviewed the chemistry of these ions[23] and the number of known examples has since then increased substantially. Cations of the allylic type are particularly stable. Representative examples which have been observed are the ions *(15)* to *(25)*.

(15)

(16)

(17)

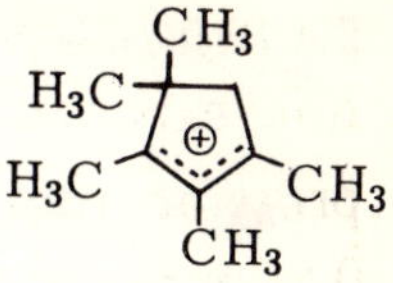

(18)

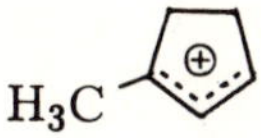

(19)

(20)

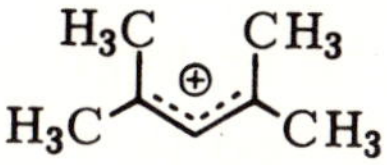

(21)

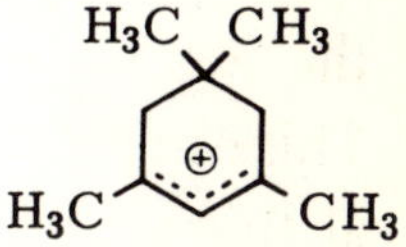

(22)

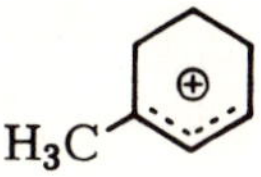

(23)

(24)

(25)

The opening of halocyclopropanes with the formation of allyl cations such as *(16)* has also been investigated[58–61].

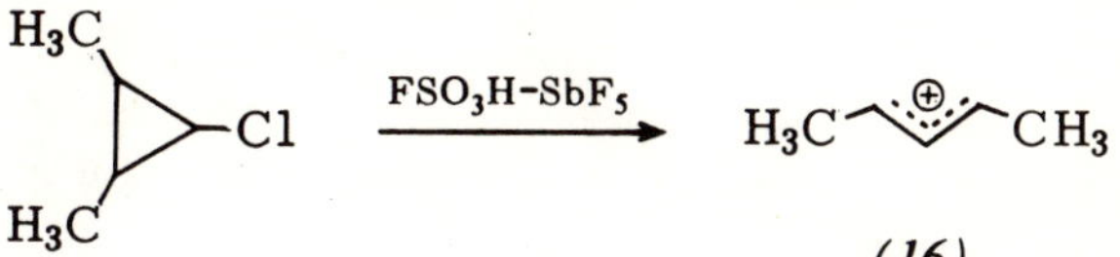

(16)

Protonation of certain allenes also leads to allyl cations which are otherwise difficult to obtain from allylic precursors[62,63].

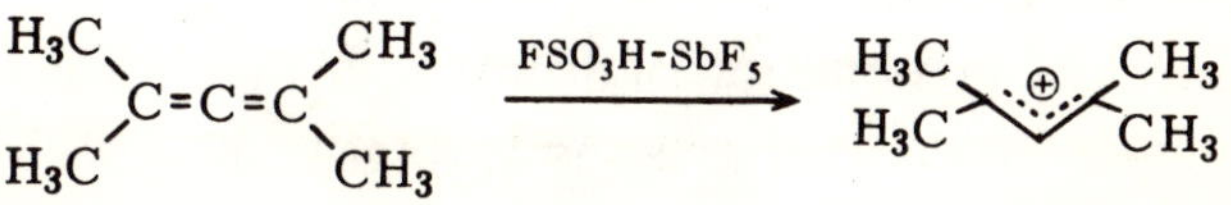

(26)

## 2.4. Alkadienyl and Polyenyl Cations

*Deno*, *Richey*, and their co-workers[23] have investigated a substantial number of alkadienyl and polyenylic cations. *Sorensen*[64] has reported the observation of the divinyl- *(27)* and trivinylcarbonium ions *(28)*.

*(27)*

*(28)*

Alkenyl cations show a substantial tendency to cyclize to the more stable cyclic allylic ions such as *(18)* and its isomers *(18a)*. *Pittman*, *Sorensen*, and *I* were able to follow these cyclizations by NMR spectroscopy[65].

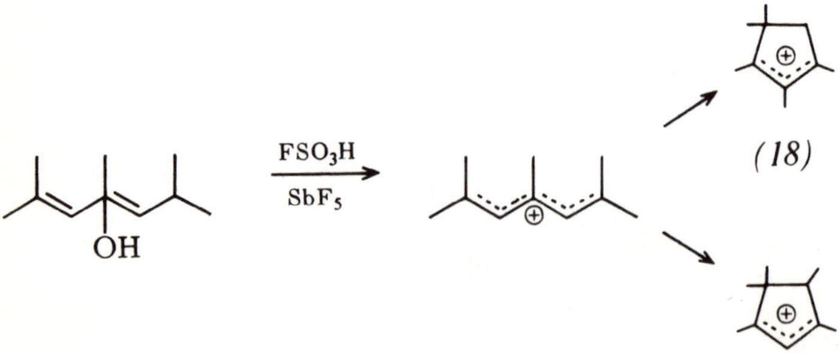


## 2.5. Arenium Ions

Cycloalkadienyl cations, particularly cyclohexadienyl cations (benzenium ions), show remarkable stability. They can be obtained by protonation of aromatic hydrocarbons by strong acids[66–68] and studied advantageously by $^1H$- or $^{13}C$-NMR spectroscopy[69].

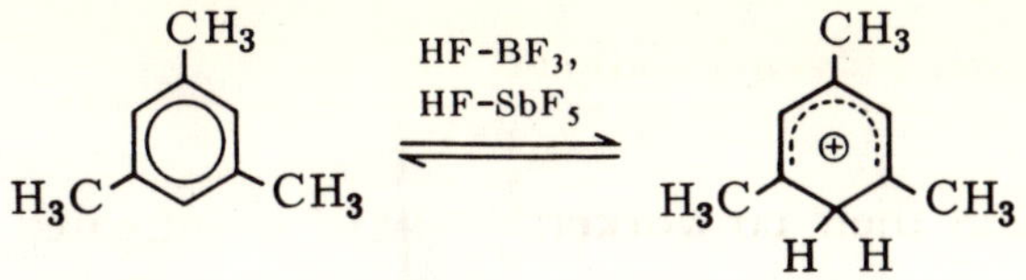


*Doering* and *Saunders* described the very stable heptamethylbenzenium ion *(29)*[70].

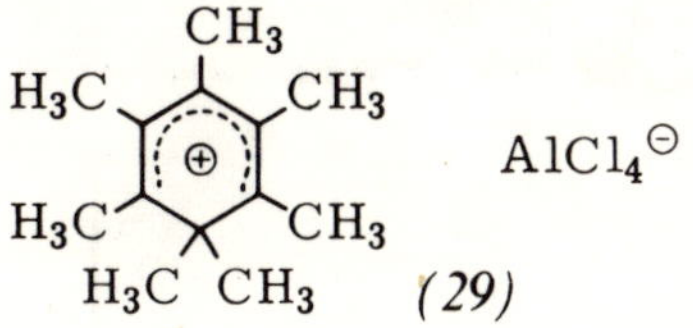


*(29)*

Figures 8, 9, and 10 show the $^1$H-NMR spectra of the *p*-toluenium ion *(30)*[69], the naphthalenium ion *(31)*[71]

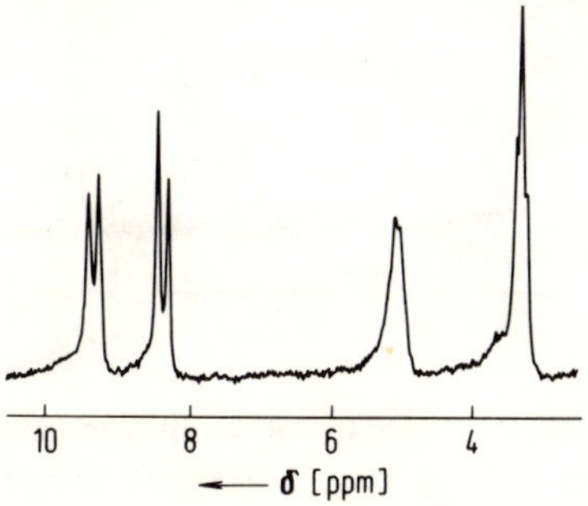


Fig. 8. 60-MHz $^1$H-NMR spectrum of the *p*-toluenium ion *(30)* at −97 °C.

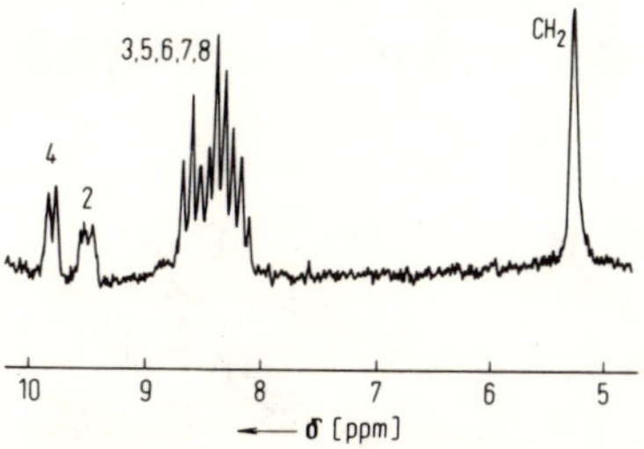


Fig. 9. 100-MHz $^1$H-NMR spectrum of the naphthalenium ion *(31)* at −90 °C.

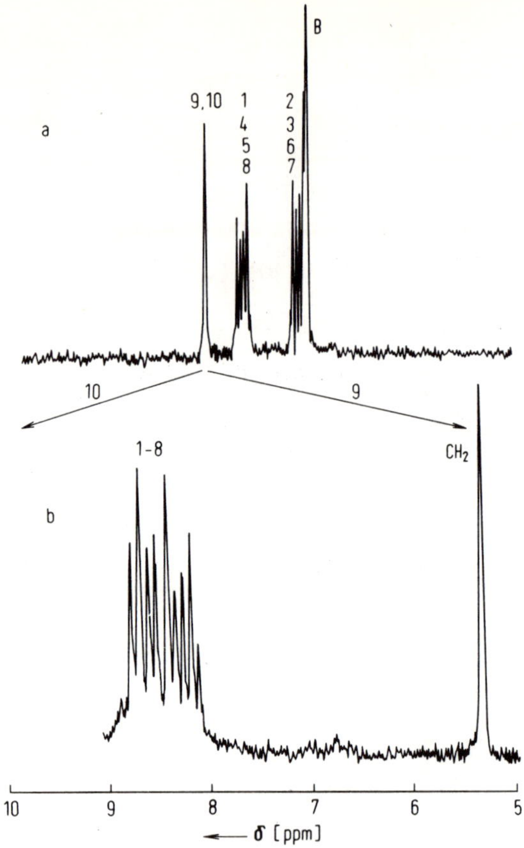


Fig. 10. (a) 100-MHz $^{1}$H-NMR spectrum (−55 °C) of anthracene in $C_6D_6$ and (b) of the anthracenium ion *(32)*, $B = C_6H_6$; (c) part of (b);

and the anthracenium ion *(32)*[72] as representative examples of arenium ions (which are also of importance as intermediates, usually called σ-complexes, in electrophilic aromatic substitutions).

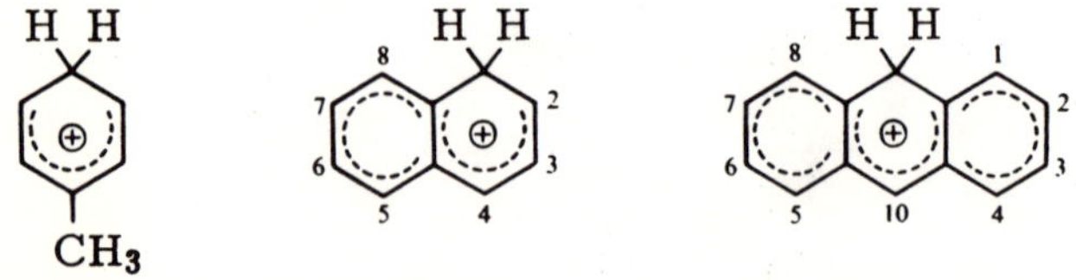

*(30)* *(31)* *(32)*

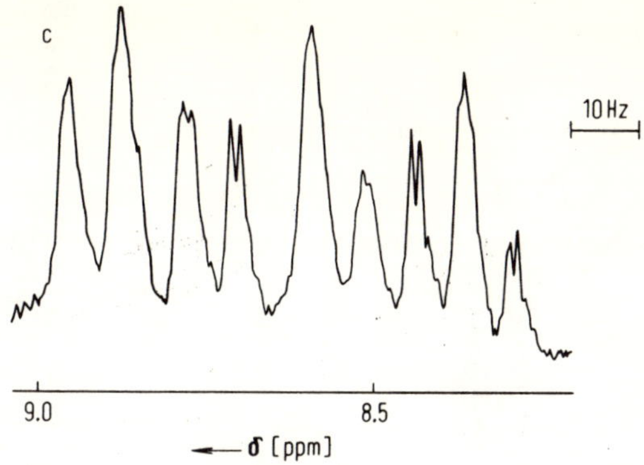


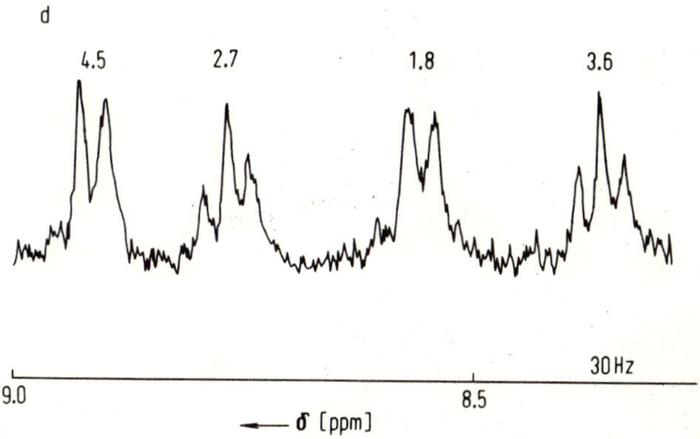


(d) first order pattern of the 300-MHz $^{1}$H-NMR spectrum of the anthracenium ion.

Even the parent benzenium ion ($C_6H_7^+$) *(33)* has been observed[73], as have also the isomeric monoalkylbenzenium ions[71, 74] and most di-, tri-, tetra- and pentaalkyl and halobenzenium ions[74]. Some representative ions are *(34)*—*(36)*.

*(33)*

*(30)*

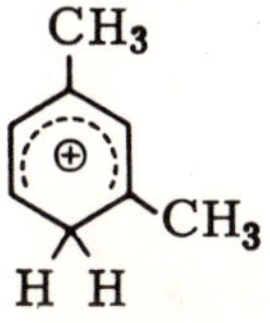
*(34)*

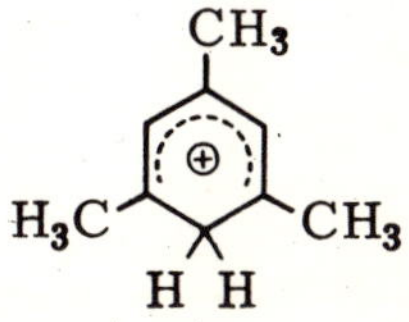
*(35)*

*(36)*

## 2.6. Cycloalkyl Cations

Tertiary cycloalkyl cations, like the 1-methyl-1-cyclopentyl cation *(37)*, show high stability in superacid solutions. The ion *(37)* can be obtained from a variety of starting substances[75].

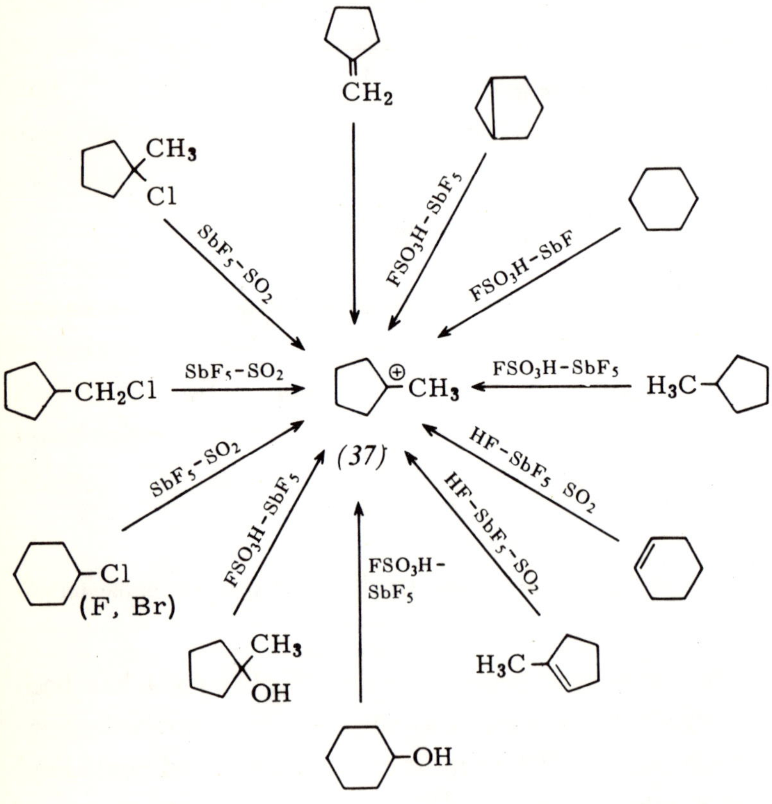


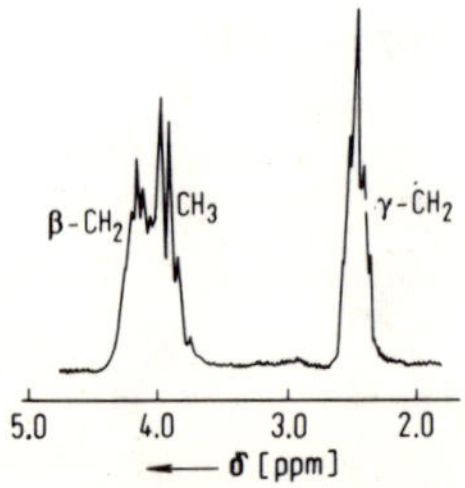


Fig. 11. 60-MHz $^1$H-NMR spectrum of the 1-methyl-1-cyclopentyl cation *(37)* at $-60\,°C$.

It is worthwhile mentioning that not only cyclopentyl but also cyclohexyl compounds give the 1-methyl-1-cyclopentyl cation. This indicates that the tertiary methyl-cyclopentyl cation is more stable than the secondary cyclohexyl cation, from which it is formed by isomerization.

The cyclopentyl cation shows in its proton magnetic resonance spectrum in $SbF_5$—$SO_2ClF$ solution, even at 100 °C, only a single absorption line at δ 4.68[38, 73]. This observation indicates a completely degenerate ion in which there are only low energy barriers to the consecutive secondary-secondary hydrogen shifts (see below).

The secondary cyclobutyl cation undergoes immediate rearrangement *via* σ-bond delocalization into the equilibrating non-classical cyclopropylmethyl ion system[76] (see Section 3.4). Similar behavior is also observed with the 1-methylcyclobutyl cation. The 1-phenylcyclobutyl cation *(38)* (Fig. 12) on the other hand is a classical tertiary carbenium ion[76].

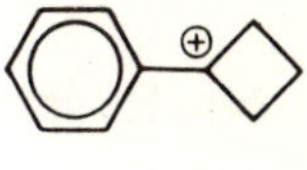

*(38)*

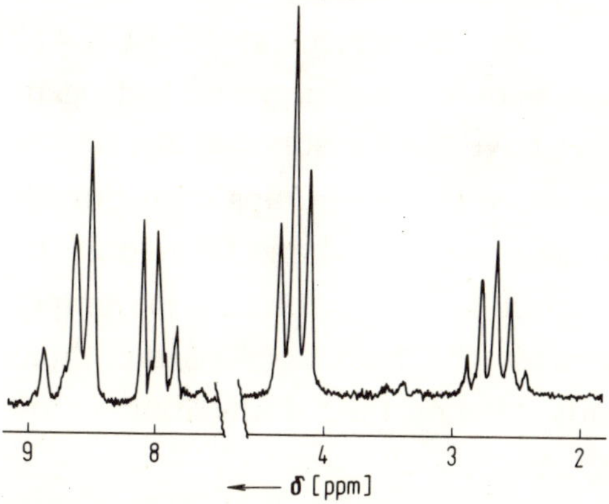


Fig. 12. 60-MHz $^1$H-NMR spectrum of the 1-phenyl-1-cyclobutyl cation *(38)* at −65 °C.

## 2.7. Bridgehead Carbenium Ions

Bredt's rule in its original form[77] seems to exclude the possibility of formation of positive carbenium ion centers at bridgehead carbon atoms of bicycloalkanes. Indeed, bridgehead halides, such as apocamphyl chloride, proved extremely unreactive under solvolytic conditions[78]. However, 1-bromoadamantane very readily gives the bridgehead carboxylic acid under the usual conditions of the Koch-Haaf acid synthesis[79]. 1-Fluoroadamantane is ionized in $SbF_5$ to give the stable bridgehead 1-adamantyl cation *(39)*[80, 81].

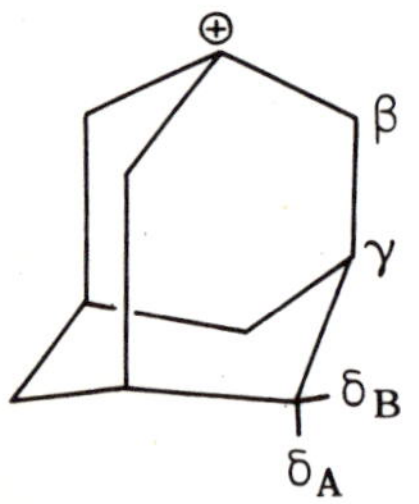


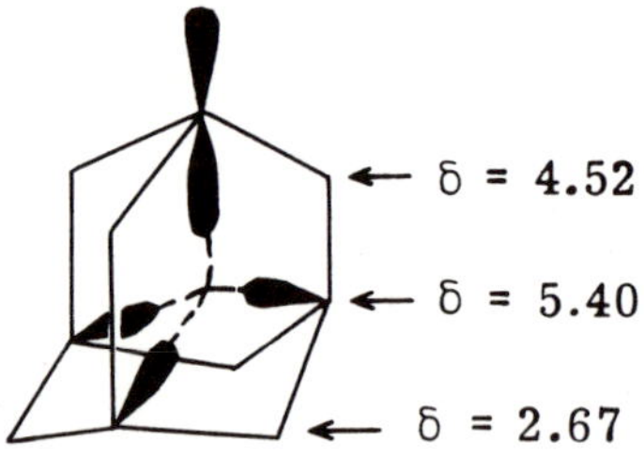


*(39)*

The $^1$H-NMR spectrum of the adamantyl cation *(39)* in $SbF_5$ solution consists of resonences at δ 5.40, δ 4.52 and δ 2.67 with peak areas of 3:6:6. The γ-CH hydrogens are more deshielded, *i.e.* have their resonance at a lower field, than those of the three β-$CH_2$ groups even though they are farther from the positive carbon. This could be explained by a "cage effect", *i.e.* the lobe of the empty p-orbital overlapping with the backlobes of the three bridgehead C—H bonds, causing their deshielding, or by σ-π hyperconjugation.

A similar effect is indicated in the $^{13}$C-NMR spectrum of the 1-adamantyl cation (Fig. 13).

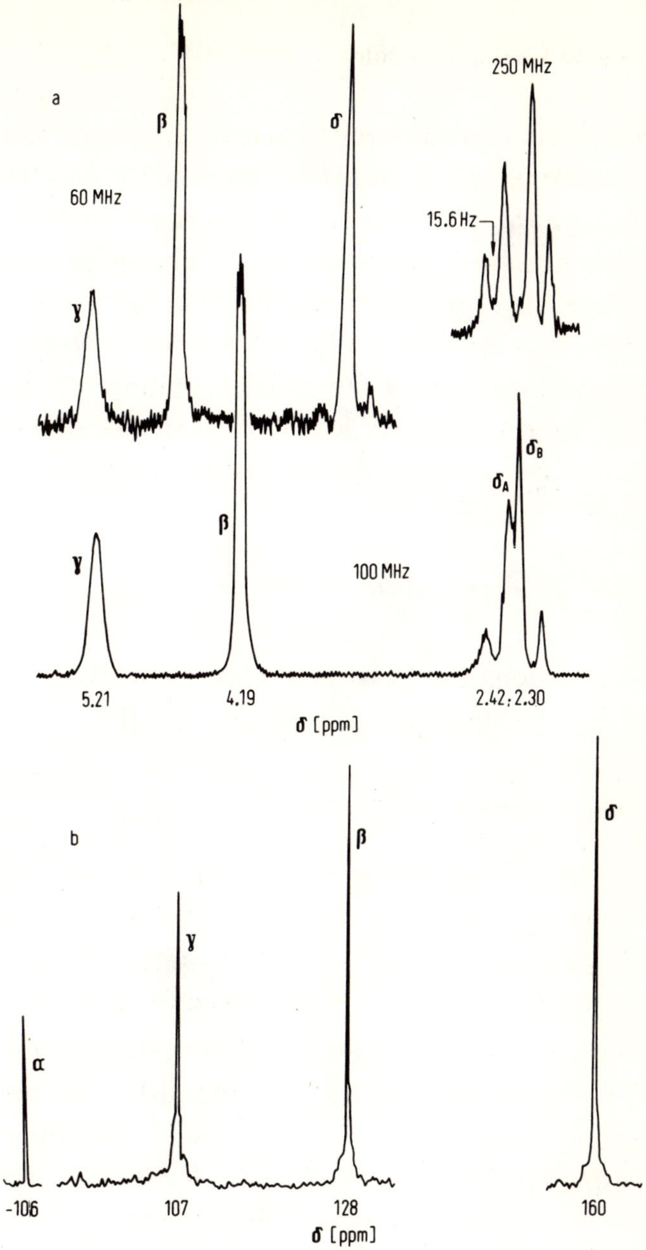


Fig. 13. a) $^1$H-NMR spectrum of the 1-adamantyl cation at 60 MHz, 100 MHz, and 250 MHz; b) Fourier-transform $^{13}$C-NMR of the 1-adamantyl cation (in $FSO_3H—SbF_5$).

*Mateescu*, *Sipos*, *Schleyer*, and *I* have also observed a series of methyladamantyl cations[81]. The bridgehead "congressane" (diadamantane) cation has also been observed recently[75,81]; it likewise shows the cation cage effect. Although the bridgehead 1-norbornyl cation could not be observed directly, the stable 2-norbornyl cation can be obtained from 1-chloronorbornane in $SbF_5$—$SO_2$ solution[82]. Thus, ionization to the bridgehead carbenium ion must be followed by a fast shift of hydrogen from C-2 to C-1, the driving force for which is obviously the tendency to relieve strain in the carbenium ion.

## 2.8. Aryl and Alkylarylcarbenium Ions

The first stable long-lived carbenium ion to be observed was the triphenylcarbenium ion *(40)*[83–85] (Fig. 14a).

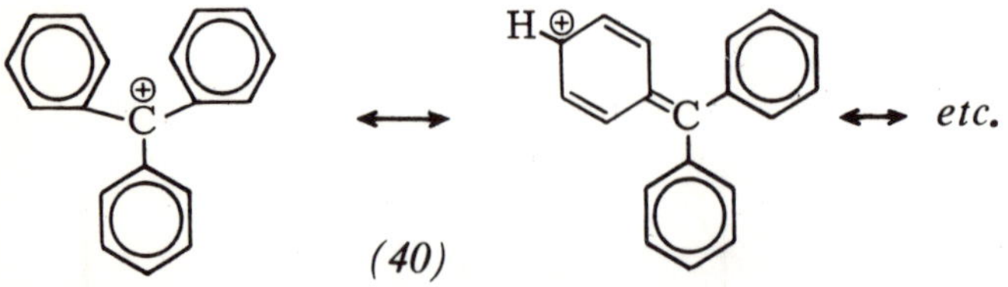


*(40)*

This ion remains the most-investigated carbocation[2] and its propeller-shaped structure has been established beyond doubt. Strong contribution from *para* (and *ortho*) quinonoidal resonance forms in the ion make it probable that these reactive secondary forms contribute strongly to the reactivity of the ion. Diphenylcarbenium ions (benzhydryl cations) are considerably less stable than their tertiary analogs. Although their ultraviolet spectra have been

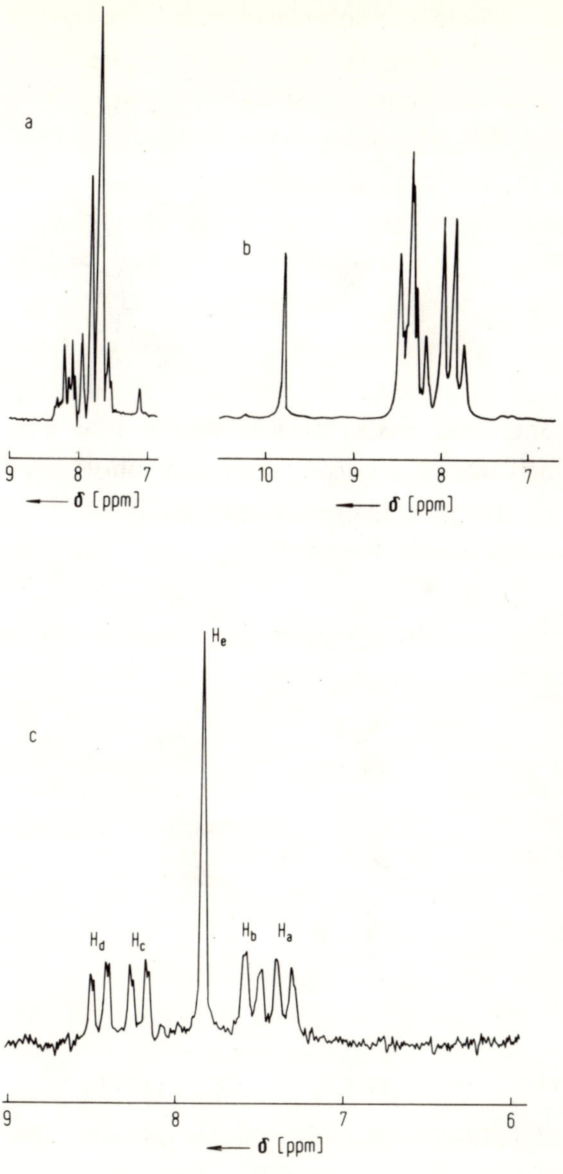


Fig. 14. a) 60-MHz $^1$H-NMR spectrum of the triphenylcarbenium ion *(40)* (trityl cation); b) 60-MHz $^1$H-NMR spectrum of the diphenylcarbenium ion *(41)* (benzhydryl cation); c) 100-MHz $^1$H-NMR spectrum of the 4-methoxyphenylcarbenium ion (*p*-methoxybenzyl cation).

obtained in dilute sulfuric acid solutions[86] the benzhydryl ion *(41)* has only recently been observed in higher concentrations in acid solutions ($ClSO_3H$[87], $FSO_3H$[88], and $FSO_3H—SbF_5$[89]) (Fig. 14b).

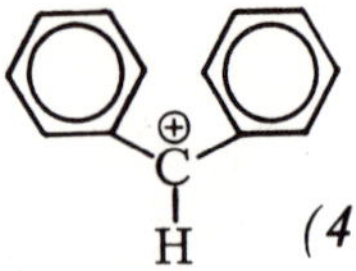

*(41)*

Solutions of mono- and dialkylarylcarbenium ions can be readily obtained from the corresponding alcohols, olefins, or halides in strong acid media such as $H_2SO_4$[86],

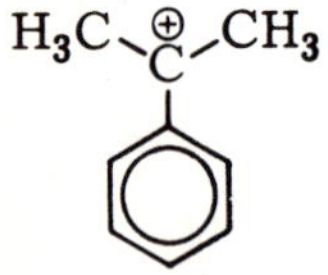

*(42)*

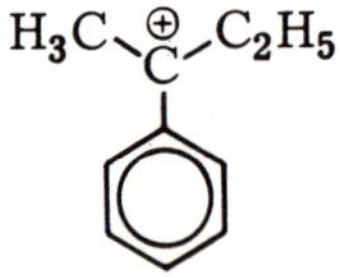

*(43)*

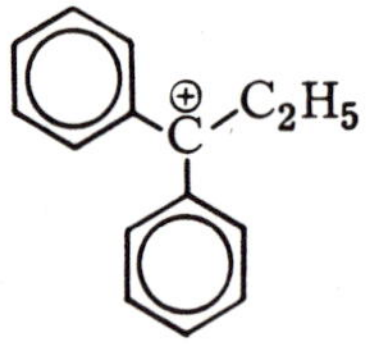

*(44)*

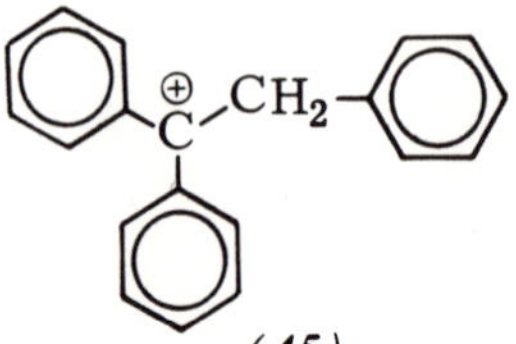

*(45)*

$FSO_3H—SbF_5$[89], $ClSO_3H$ and $FSO_3H$[87], and oleum[88]. Representative examples of the cations are *(42)*—*(45)*.

Because of the high stability of the tertiary ions, these are preferentially formed in the strong acid systems from both tertiary, secondary, and even primary precursors[90].

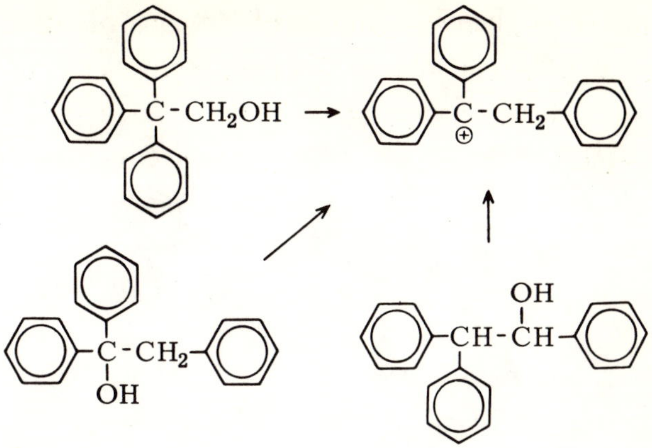


Non-benzylic, tertiary arylalkylcarbenium ions, on the other hand, rearrange to secondary, benzylic ions such as *(46)*.

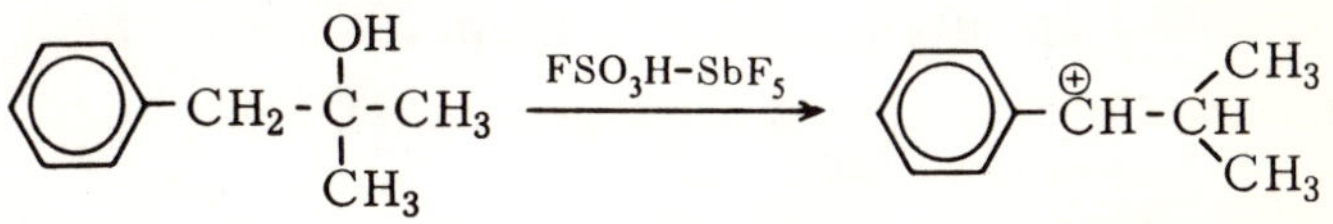


*(46)*

Suitable substituent groups (which also prevent transalkylations) stabilize even secondary [*e.g.* *(47)*] and primary benzyl cations [*e.g.* *(48)*[91,92] and the *p*-methoxybenzyl cation, Fig. 14c].

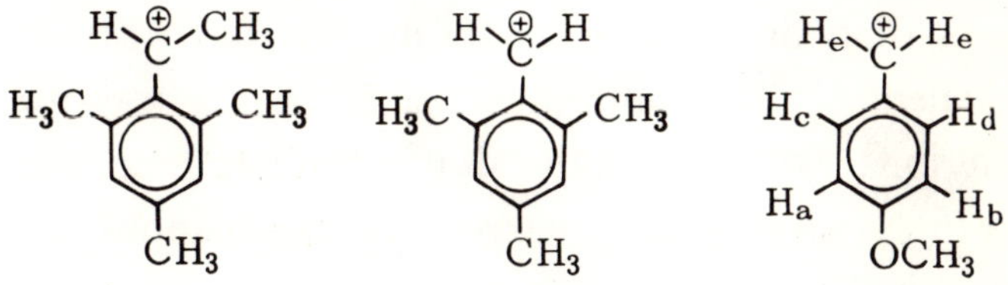


*(47)* *(48)* *(49)*

The observation of benzyl cations in superacid solutions was only recently achieved[91]. Although the unsubstituted benzyl cation is still elusive, we have observed a number of substituted derivatives such as *(50)—(53)*.

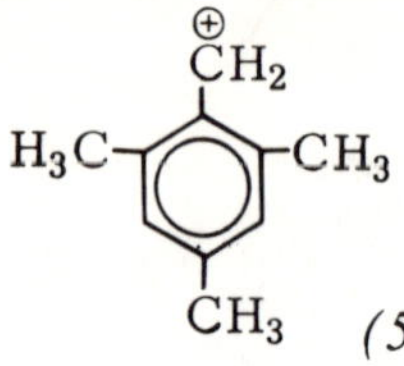

*(50)*

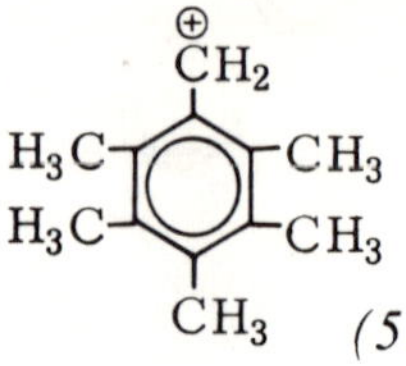

*(51)*

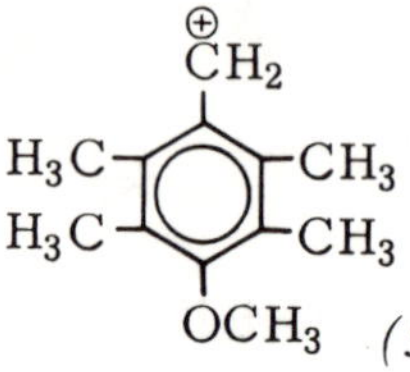

*(52)*

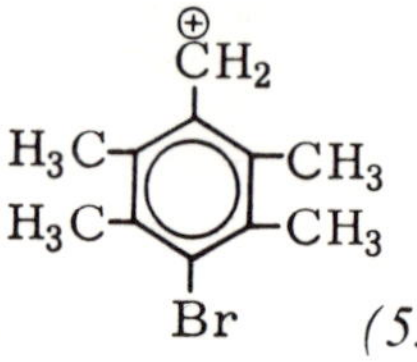

*(53)*

As NMR studies have shown, in benzyl cations like 2,4-dimethyl-6-*tert*-butylphenylcarbenium ion *(54)* a high rotational barrier is observed around the $H_2^+C—C_1$ bond.

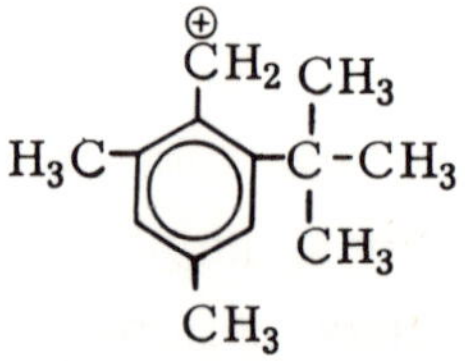

*(54)*

This means that the methylene protons are magnetically non-equivalent[92]. Rearrangement of benzyl cations to tropylium ions, which is claimed to take place in the gas phase (mass spectroscopic measurements) has not been observed in acid solution.

The styryl cation (methylphenylcarbenium ion) *(55)*[93] and a number of substituted methylphenylcarbenium ions *(56)—(58)*[85] have also been observed.

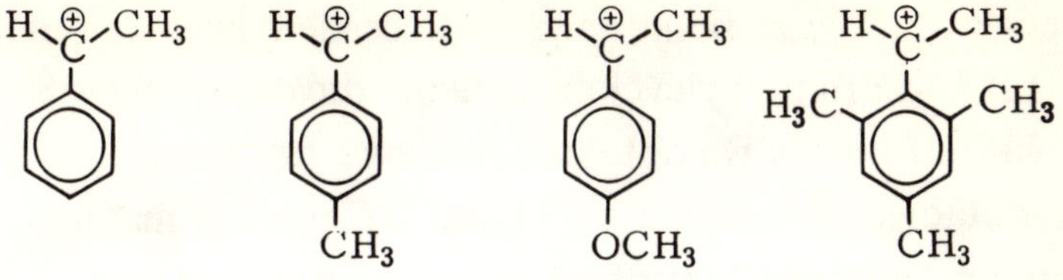


(55) (56) (57) (58)

## 2.9. Dicarbenium Ions

Earlier reports[94] that a dicarbenium ion had been obtained from pentamethyltrichloromethylbenzene have turned out to be incorrect. The species in question is the benzylic dichloropentamethylphenylcarbenium ion *(59)*[95-97].

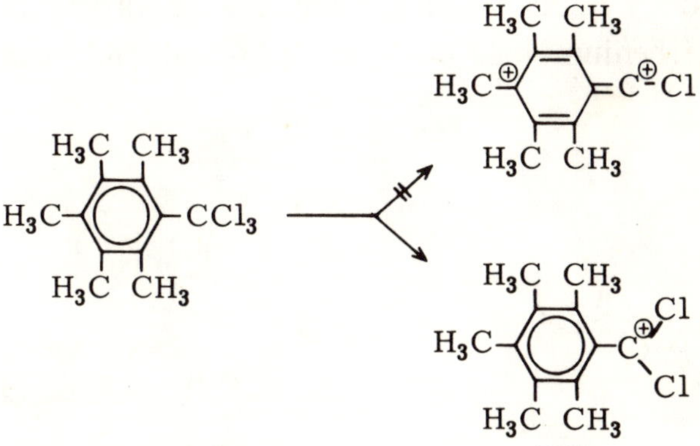


(59)

The 1,2,3,4-tetraphenylcyclobutadiene dication has also been reported[98]. However, there is evidence that the ion actually observed was only an equilibrating cyclobutenyl monocation. X-ray crystallographic investigation of the isolated crystalline complex confirm this conclusion[99].

Recently *Bollinger*, *White*, and *I* succeeded in obtaining the 1,2,3,4-tetramethylcyclobutadiene dication *(60)*[100] and with *Mateescu* we obtained the *bona fide* tetraphenylcyclobutadiene dication[101]. $^1H$ and $^{13}C$ nuclear magnetic resonance studies confirmed its structure and the 2π electron aromatic nature, which is similar to the cyclopropenium cations previously prepared by *Breslow*[102].

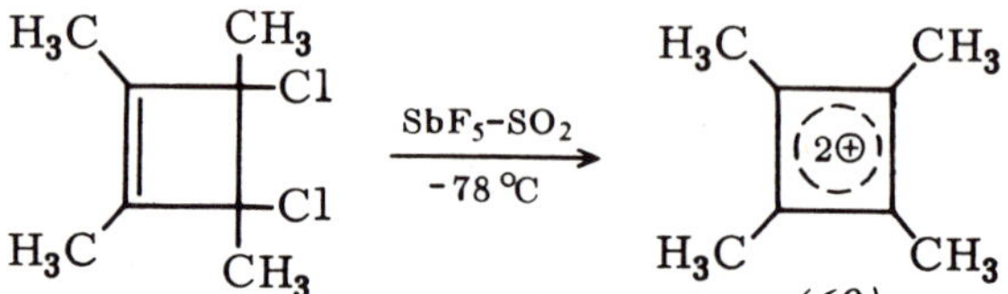

*(60)*

*Hart* and *Volz* have shown that if two carbenium centers are separated by a phenyl ring, then a wide variety of di- and tricarbenium ions such as *(61)* and *(62)* can be obtained[103, 104].

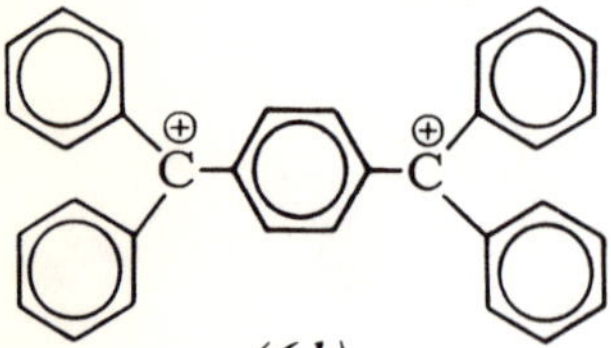
*(61)*

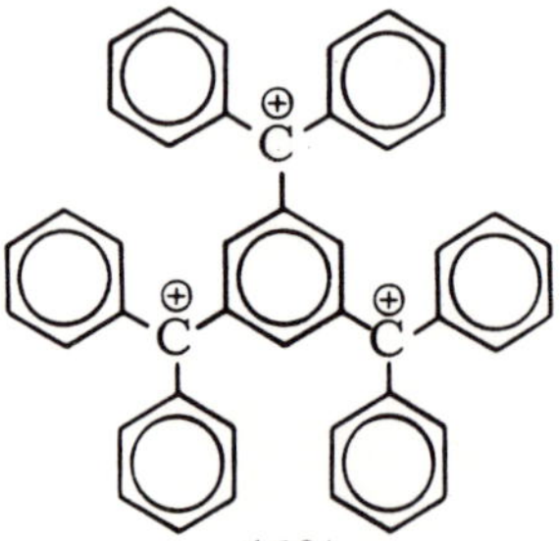
*(62)*

We have also prepared simple alkyldicarbenium ions[105], in which the two carbenium ion centers are separated by two or three methylene groups. These ions [*(63)* and

*(64)*, Fig. 15] are stable entities in the strongly acidic solvent system ($SbF_5$—$SO_2$) at low temperature (−60 °C).

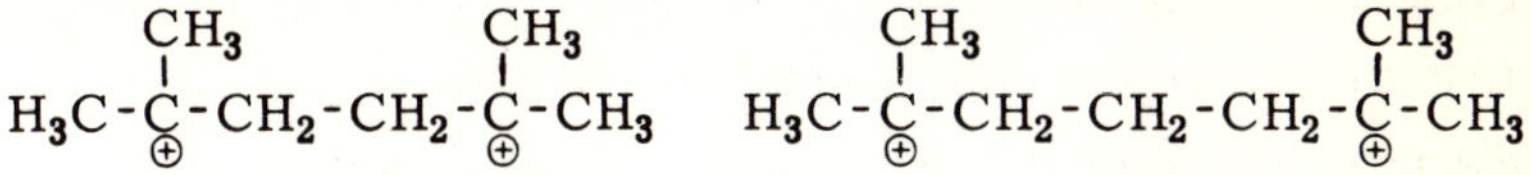

*(63)* *(64)*

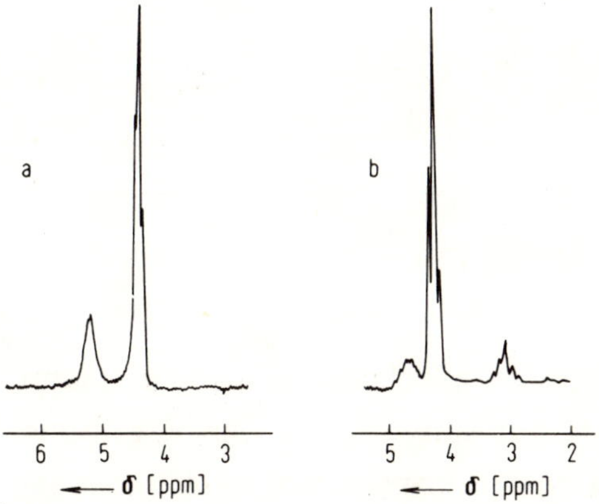


Fig. 15. 60-MHz $^1$H-NMR spectra of the alkyldicarbenium ions, a) *(63)* and b) *(64)*.

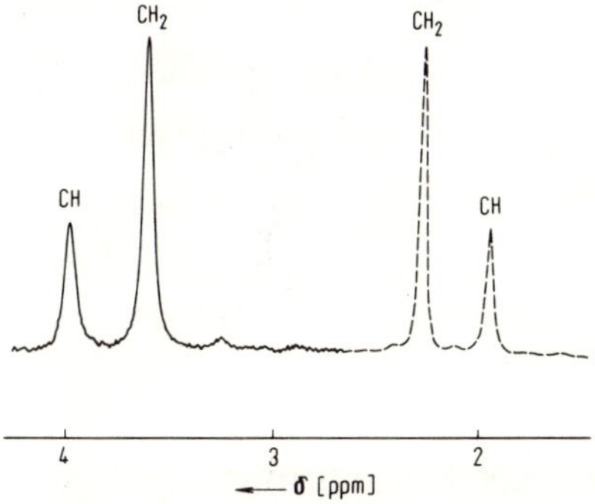


Fig. 16. 100-MHz $^1$H-NMR spectrum of the apical,apical diadamantane dication *(65)* (———) and the precursor dibromoadamantane *(66)* (– – –).

Dicarbenium ions have also been observed in more rigid cage systems, such as the apical,apical diadamantane dication *(65)*[81] (Fig. 16).

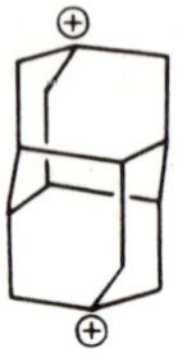

*(65)*

*(66)*

## 2.10. Aromatically Stabilized Carbenium Ions

If a carbenium ion is also a Hückeloid aromatic system, resonance causes substantial stabilization. The best known examples of aromatically stabilized carbenium ions are the cycloheptatrienyl cations *(67)* (tropylium ions)[106–108], cyclopropenium ions[102], and the tetramethyl- and tetraphenylcyclobutenium dications[100, 101].

*(67)*

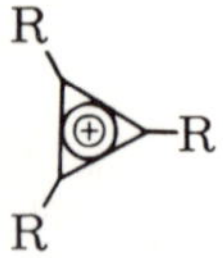


*(68)*

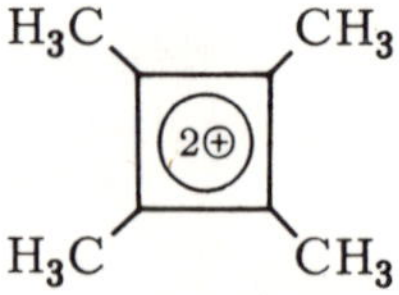


*(60)*

Because of their aromatic character, these ions are highly stable. Other examples of aromatically stabilized carbenium ions are also known.

## 2.11. Heteroatom-Stabilized Carbenium Ions

In contrast to hydrocarbon cations, heteroatom-substituted carbenium ions are strongly stabilized by electron donation from the non-bonded electron pairs of the heteroatoms adjacent to the cationic carbon center:

$$R_2\overset{\oplus}{C}-\ddot{X} \leftrightarrow R_2C=X^{\oplus} \qquad X = \mathrm{Br, O, N, S, F, Cl}$$

The stabilizing effect is enhanced when two, or even three, electron-donating heteroatoms coordinate with the electron-deficient carbon atom.

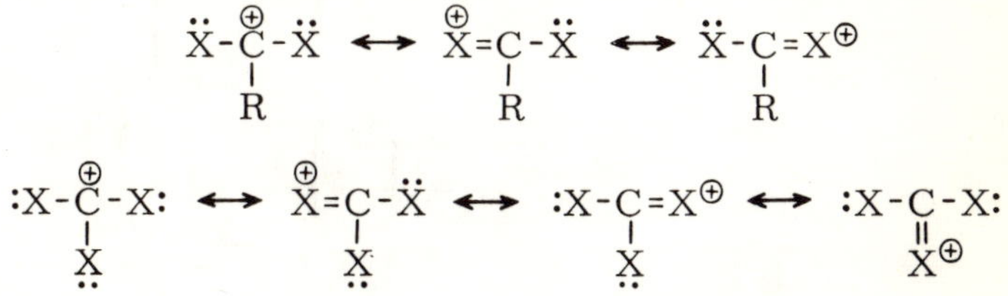

## 2.12. Halogen as Heteroatom

In 1965, *Cupas*, *Comisarow*, and *I* reported the first fluorocarbenium ions[109], the phenyldifluorocarbenium ion *(69)* and the diphenylfluorocarbenium ion *(70)*.

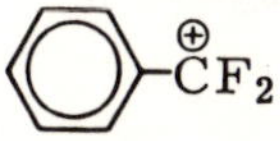

*(69)*

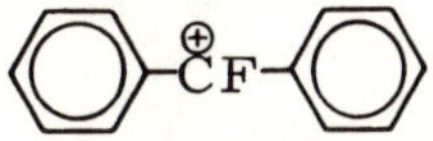

*(70)*

Fluorine has a particular ability to stabilize carbenium ions *via* back-coordination of its unshared electron pairs into the vacant p orbital of the carbon atom. Since $^{19}F$ magnetic resonance offers an excellent possibility for structural investigations of these ions, extensive work was initiated[110] in this regard. The dimethyl- *(71)* and phenylmethylfluorocarbenium ions *(72)* were *inter alia* prepared and thoroughly investigated[111].

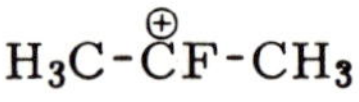

*(71)*

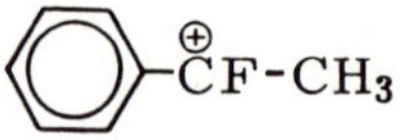

*(72)*

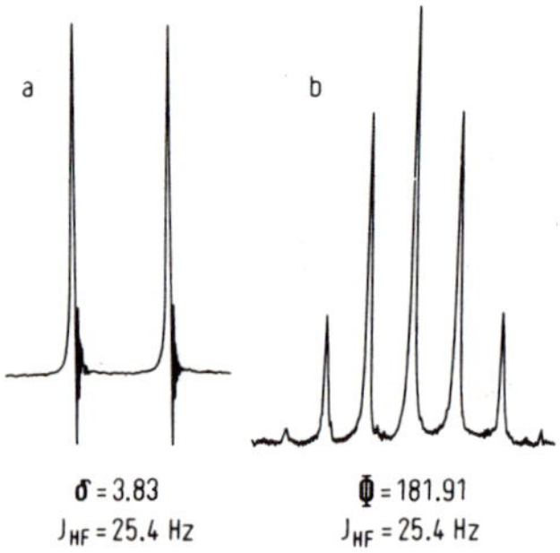


Fig. 17. a) $^{1}H$-NMR spectrum of the dimethylfluorocarbenium ion *(71)* at 60 MHz: $J_{HF} = 25.4$ Hz; b) $^{19}F$-NMR spectrum of the same ion at 56.4 MHz: $J_{HF} = 25.4$ Hz.

Trifluoromethylcarbenium ions such as *(73)*[112] and perfluorophenylcarbenium ions *(74)*—*(76)*[113,114] have also been investigated. The trifluorocyclopropenium ion *(77)*[115] is also known.

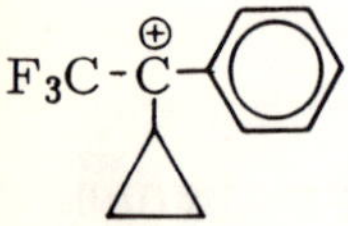

*(73)*

$(F_5C_6)_3C^{\oplus}$ *(74)*

$(F_5C_6)_2\overset{\oplus}{C}H$ *(75)*

$F_5C_6-\overset{\oplus}{C}H_2$ *(76)*

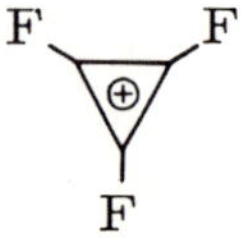

*(77)*

Because of the relatively large fluorine resonance shifts, anisotropy and ring current effects play a relatively much smaller role in $^{19}$F-NMR spectroscopy than they do in the case of proton resonance. Therefore, as is also true of $^{13}$C-NMR spectroscopy, a better correlation between charge distribution and chemical shifts can be obtained than with $^{1}$H-NMR spectra.

A series of chlorocarbenium ions, including phenyldichlorocarbenium ions such as *(59)*, *(78)* and *(79)*[95-97,116] and the perchlorotriphenylcarbenium ion *(80)*[117] *(Ballester)* have also been described. *West* observed the perchloroallyl cation *(81)*[118].

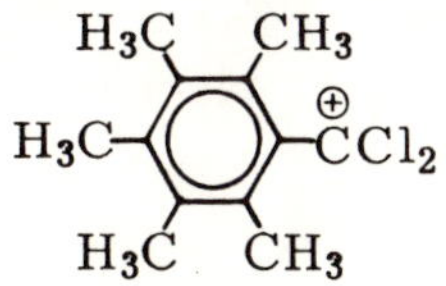

*(59)*

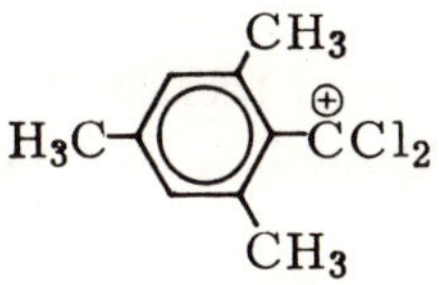

*(78)*

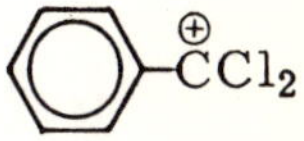

*(79)*

$(Cl_5C_6)_3C^{\oplus}$

*(80)*

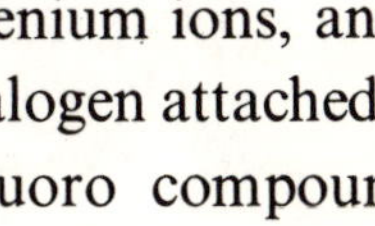

*(81)*

*Comisarow* and *I* observed[119] a series of chloro-, as well as bromo- and iodocarbenium ions, and showed the general stabilizing ability of halogen attached to carbenium ion centers. [See also the fluoro compounds *(71)* and *(72)*.] *Halpern*, *Mo*, and *I* were subsequently able to study these effects in more detail using $^{13}$C-NMR spectroscopy[120].

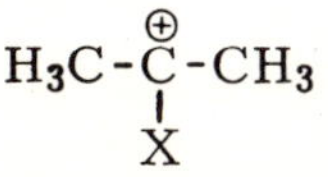

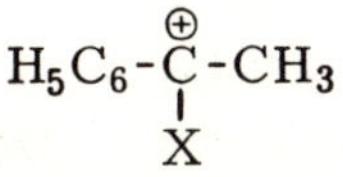

X = I, Br, Cl, F

Closely related to the halocarbenium ions are the halonium ions in which the positive charge is to be found predominantly on the halogen atom. In our work with *Bollinger*[121] and *Peterson*[122], we reported on the three- and five-membered halonium ions of the type *(82)* and *(83)*.

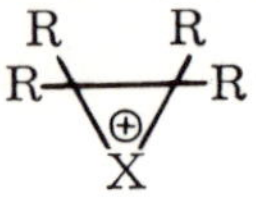

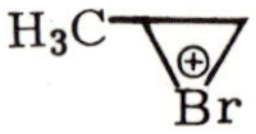


*(82)* *(83)* *(82a)*

X = I, Br, Cl

R = H or Alkyl

Illustrative is the $^1$H-NMR spectrum of propylenebromonium ion *(82a)* shown in Fig. 18.

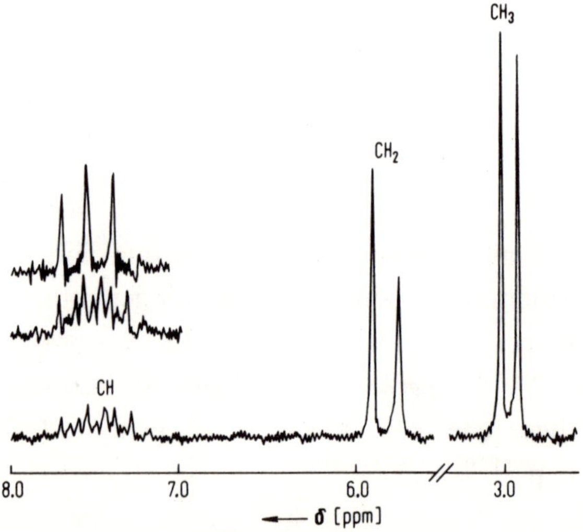


Fig. 18. $^1$H-NMR spectrum (at 60 MHz) of the propylenebromonium ion *(82a)*. Left: Signal of the methine protons, decoupled.

In our work with *DeMember*[123] we were able to obtain not only ring halonium ions, but also a series of open-chain dialkylhalonium ions such as:

| | | |
|---|---|---|
| $[H_3C{-}I{-}CH_3]^{\oplus}$ | $[[(CH_3)_2CH]_2Br]^{\oplus}$ | $[H_3C{-}I{-}C_2H_5]^{\oplus}$ |
| $[H_3C{-}Br{-}CH_3]^{\oplus}$ | $[[(CH_3)_2CH]_2Cl]^{\oplus}$ | $[H_3C{-}Br{-}C_2H_5]^{\oplus}$ |
| $[H_3C{-}Cl{-}CH_3]^{\oplus}$ | $[[(CH_3)_2CH]_2I]^{\oplus}$ | $[H_3C{-}Cl{-}C_2H_5]^{\oplus}$ |
| $[(C_2H_5)_2Br]^{\oplus}$ | $[(n\text{-}C_3H_7)_2I]^{\oplus}$ | $[CH_3{-}Cl{-}CH(CH_3)_2]^{\oplus}$ |
| $[(C_2H_5)_2Cl]^{\oplus}$ | $[(n\text{-}C_3H_7)_2Br]^{\oplus}$ | |
| $[(C_2H_5)_2I]^{\oplus}$ | | |

Dialkylhalonium fluoroantimonate salts can be isolated; they are versatile, very reactive alkylating agents, surpassing in many cases the selectivity and utility of *Meerwein*'s trialkyloxonium salts.

## 2.13. Oxygen as Heteroatom

Crystalline addition compounds of alcohols, ethers, aldehydes, and ketones with Brønsted and Lewis acids have been known since the middle of the last century. They were long considered as unstable "molecular compounds"[124]. *Collie* and *Tickle*[125] were the first (1899) to assign "oxonium salt" character to the acid complexes, *i. e.* they considered the compounds as containing a tetravalent oxygen, analogous to the ammonium salts in which nitrogen at that time was assumed to be pentavalent.

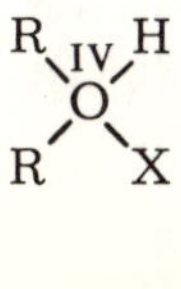


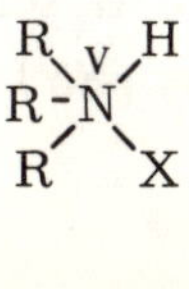

*Hantzsch*[126] proved, as early as 1922, that the pyrylium salts were clearly ionic. Pyrylium salts react with strong nucleophiles to give phenols. This behavior toward strong nucleophiles shows that the carbenium ion form contributes significantly to charge delocalization in pyrylium salts.

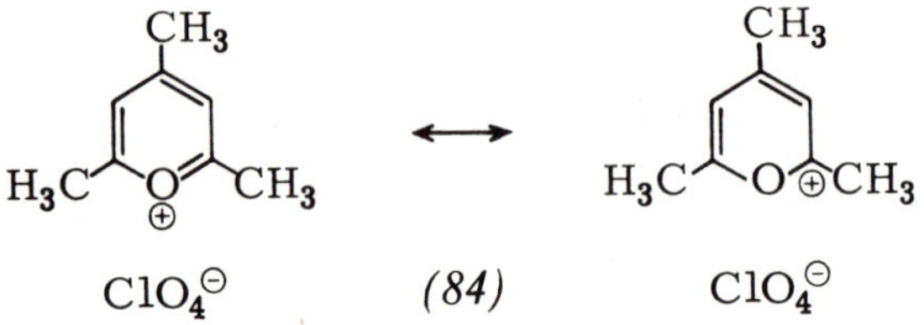

*(84)*

2.13.1. Alkoxy- and Hydroxycarbenium Ions

Resonance, similar to that in the pyrylium salts, was shown by *Meerwein*[127] to exist between tri- and dialkyloxonium and alkoxycarbenium ion forms in the fluoroborates obtained by alkylation of ketones, esters, and lactones with trimethyl- or triethyloxonium fluoroborates, such as *(85)*.

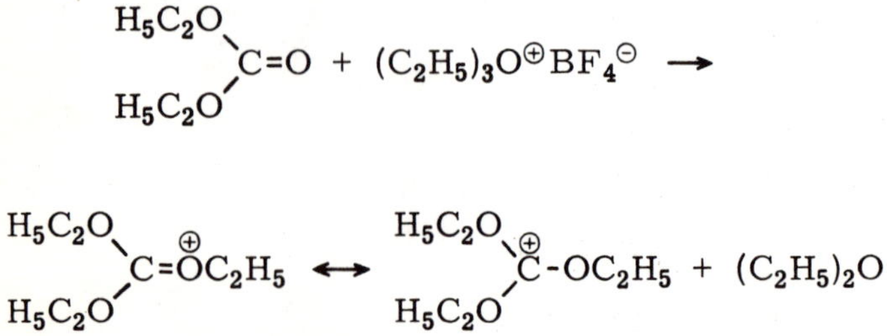

*(85)*

*Taft* and *Ramsey*[128] investigated a series of secondary and tertiary alkoxycarbenium ions *(86)*—*(88)* $^{1}$H-NMR spectroscopically.

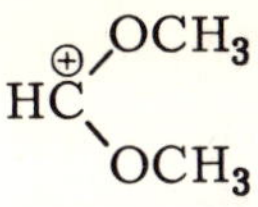

*(86)*

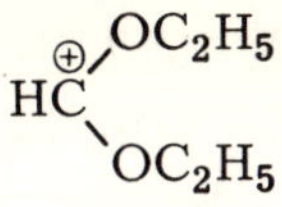

*(87)*

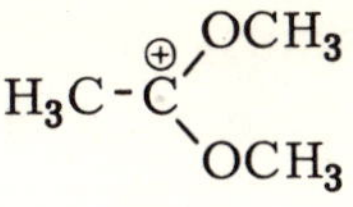

*(88)*

*Bollinger*[129] and *I* prepared primary alkoxycarbenium ions such as the methoxy- and phenoxycarbenium ions and their halogenated derivatives, and investigated their $^{1}H$-, $^{13}C$-, and later $^{17}O$-NMR spectra.

$$CH_3O\overset{\oplus}{C}H_2 \longleftrightarrow CH_3\overset{\oplus}{O}{=}CH_2$$

$$CH_3O\overset{\oplus}{C}HCl \longleftrightarrow CH_3\overset{\oplus}{O}{=}CHCl$$

$$ClCH_2O\overset{\oplus}{C}H_2 \longleftrightarrow ClCH_2\overset{\oplus}{O}{=}CH_2$$

$$CH_3O\overset{\oplus}{C}HF \longleftrightarrow CH_3\overset{\oplus}{O}{=}CHF$$

$$C_6H_5O\overset{\oplus}{C}H_2 \longleftrightarrow C_6H_5\overset{\oplus}{O}{=}CH_2$$

Acidic oxonium ions can be studied particularly well in superacid solvent systems. With *Sommer* and *Namanworth*[41] we showed that primary and secondary alcohols are protonated in $FSO_3H$—$SbF_5$($SO_2$, $SO_2ClF$) solution at $-60\,^{\circ}C$, and that well-resolved NMR spectra of the protonated species can be obtained.

At higher temperatures they cleave to carbenium ions; the kinetics of these cleavage reactions could be followed by NMR spectroscopy.

$$R\overset{\oplus}{O}H_2 \xrightarrow[H^{\oplus}]{\Delta} R^{\oplus} + H_3O^{\oplus}$$

Tertiary alcohols (with the exception of the ones containing strongly electron-withdrawing groups like $CF_3$) generally dehydrate very fast in the acid media, and the intermediate protonated species cannot be observed, even at low temperature.

Ethers are also protonated in superacid media[43].

$$R_2O \xrightarrow[-60\,^{\circ}C]{FSO_3H\text{-}SbF_5\text{-}SO_2} R_2\overset{\oplus}{O}H$$

Once again the subsequent cleavage reactions can, as in the case of methyl *n*-butyl ether *(7)*, be followed by NMR spectroscopy.

$$H_3C\underset{\displaystyle H}{\underset{|}{\overset{\oplus}{O}}}CH_2\text{-}CH_2\text{-}CH_2\text{-}CH_3 \xrightarrow[H^{\oplus}]{\Delta}$$

$$H_3C\overset{\oplus}{O}H_2 + [H_3C\text{-}CH_2\text{-}CH_2\text{-}CH_2^{\oplus}] \longrightarrow (CH_3)_3C^{\oplus}$$

Aldehydes and ketones are protonated on the carbonyl oxygen atom, and in superacid media at low temperatures the protonated species can be directly observed[130–134].

$$R\text{-}CHO \xrightarrow{FSO_3H\text{-}SbF_5\text{-}SO_2} R\text{-}CH{=}\overset{\oplus}{O}H$$

$$R_2CO \longrightarrow R_2C{=}\overset{\oplus}{O}H$$

Even protonated formaldehyde has been prepared. Protonated acetaldehyde was observed in two isomeric forms, the proton on oxygen being *syn* or *anti* to the hydrogen of the aldehyde group.

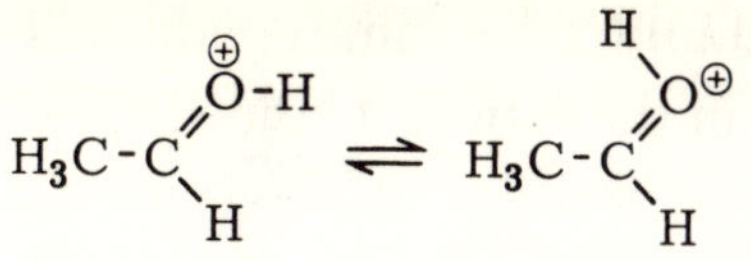

*syn*, 80% *anti*, 20%

The hydroxycarbenium ion forms of protonated ketones and aldehydes (obtained in strong acid solutions) contribute to the resonance hybrid, the degree of contribution of each form being quite accurately estimable by $^{13}C$-NMR spectroscopy[135].

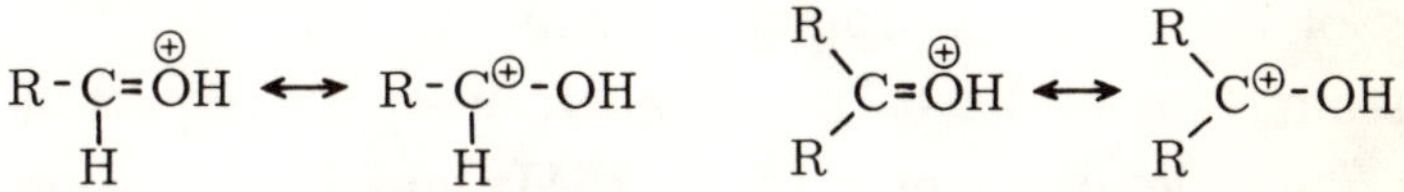

Carboxylic acids are protonated in superacid media such as $FSO_3H—SbF_5—SO_2$, $HF—SbF_5$, or $HF—BF_3$[136].

The NMR spectrum of acetic acid in such media at low temperatures shows two OH resonances. This indicates (i) that protonation of the carbonyl group is favored, and (ii) that rotation about the resultant C ⎓ OH bonds is hindered. The predominate conformer observed is the *syn,anti* isomer; but about 5 percent of the *syn,syn* isomer has also been observed.

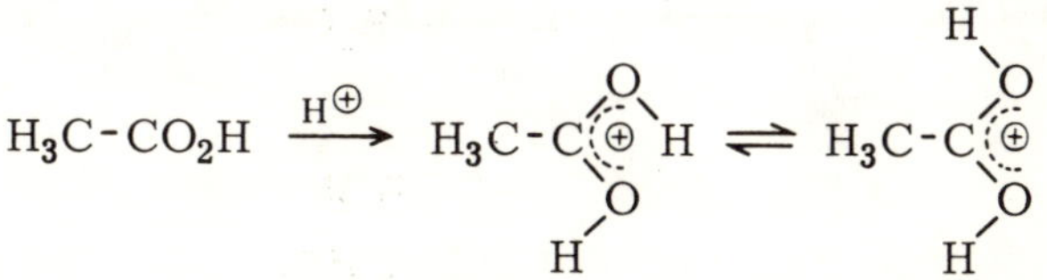

*syn, anti* *syn, syn*

95% 5%

Formic acid is also protonated under the same conditions, two conformers being formed in the ratio 2:1; the *syn,anti* isomer predominates.

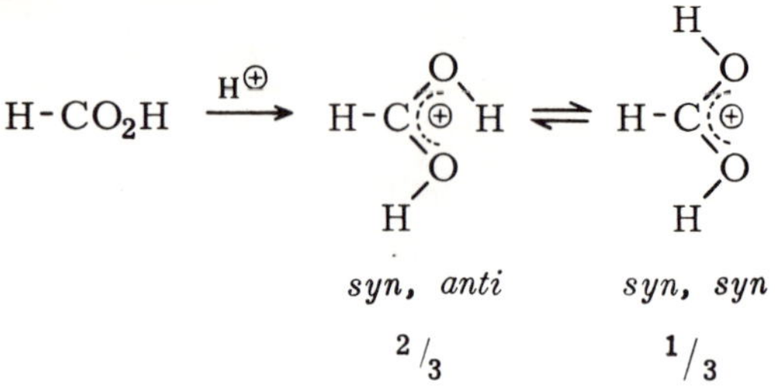


These isomers can be readily identified from the magnitudes of the vicinal coupling constants in the $^1$H-NMR spectrum; thus in the *syn,anti* isomer, the methine proton gives a doublet of doublets ($J_{HH}$ = 15 Hz and 3.5 Hz) while in the case of the *syn,syn* isomer a triplet is observed ($J_{HH}$ = 3.5 Hz). No evidence for the *anti,anti* isomer has been found in either protonated carboxylic acids, esters, or their thio analogs.

Esters behave in an analogous fashion. Once again the oxygen atom of the carbonyl oxygen takes place. Thus, protonated methyl formate is present in $FSO_3H$—$SbF_5$—$SO_2$ solution as two isomers in a ratio of 90 to 10[137].

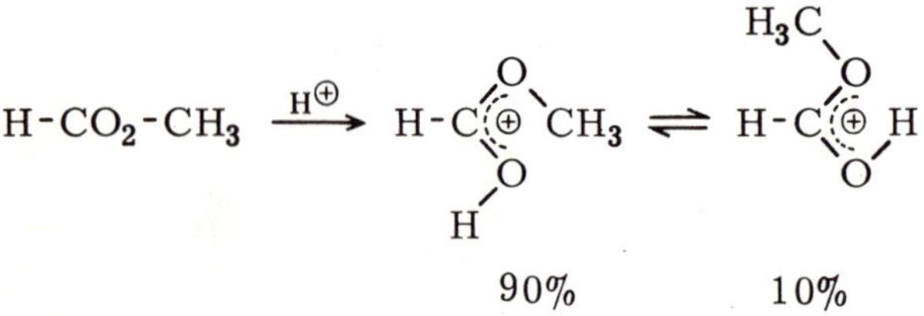


By raising the temperature of solutions of protonated carboxylic acids and esters, their cleavage reactions can be observed. These reactions can be considered within the framework of the two unimolecular reaction pathways for acid-catalyzed hydrolyses of esters, either involving alkyl- or acyl-oxygen cleavage. Studies of these reactions in superacid media have the advantage over those carried out under the usual solvolytic conditions in that the cleavage step can be isolated and studied in detail, since the cleavage products generally do not undergo further reaction.

For example, in the case of protonated acetic acid in $FSO_3H—SbF_5—SO_2$ solution, a reaction analogous to the rate-determining step in the unimolecular cleavage of esters is observed; this leads to formation of the acetyl cation *(89)* and the hydronium ion.

$$[H_3C\text{-}CO_2H_2]^{\oplus} \xrightarrow[H^{\oplus}]{-20\,^{\circ}C} H_3C\text{-}\overset{\oplus}{C}{=}O + H_3O^{\oplus}$$

*(89)*

Unimolecular cleavage in this case corresponds to dehydration of the acid. In the case of protonated esters, however, the cleavage pathway depends on the nature of the alkoxy group, as is shown in examples *(90)*—*(92)*.

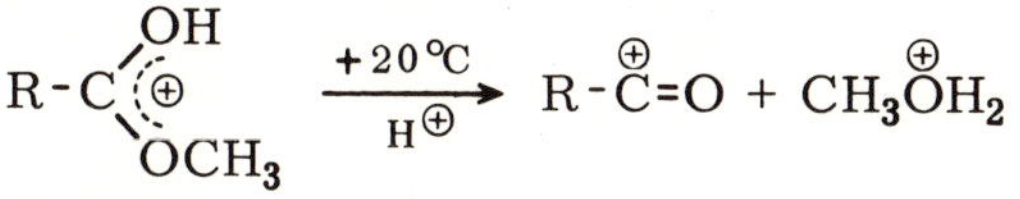


*(90)*

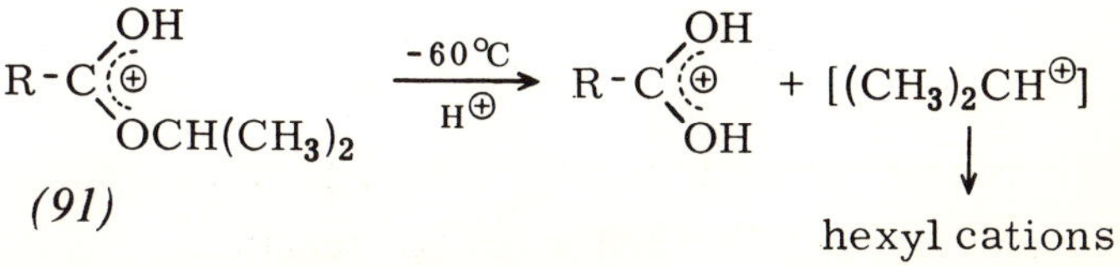


*(91)*

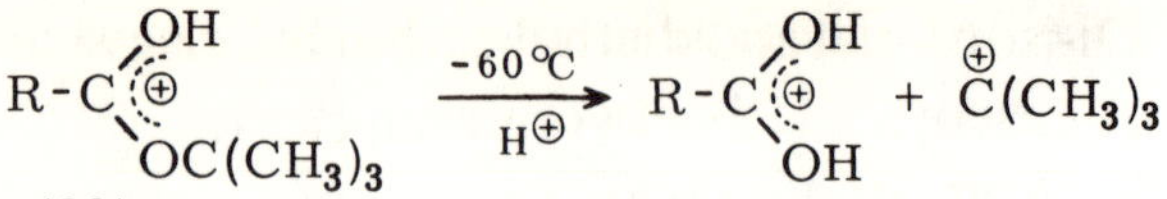

(92)

Dialkyl carbonates have been studied in $FSO_3H—SbF_5$ solution and have been shown to be protonated on the carbonyl oxygen giving the dialkoxyhydroxycarbenium ions *(93)*[138].

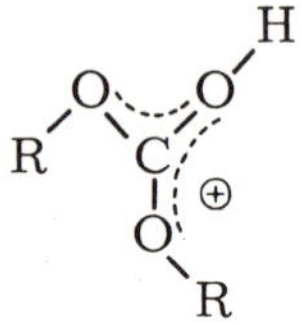

(93)

Di-*tert*-butyl carbonate cleaves spontaneously at −80 °C. Alkyl-oxygen fission takes place giving the *tert*-butyl cation and the protonated form of carbonic acid *(94)*, whose structure has been established from the $^{13}C$-NMR spectrum (see Fig. 19). The quartet with a coupling constant of 4.5 Hz shows coupling of the carbon to three equivalent hydroxyl protons[138].

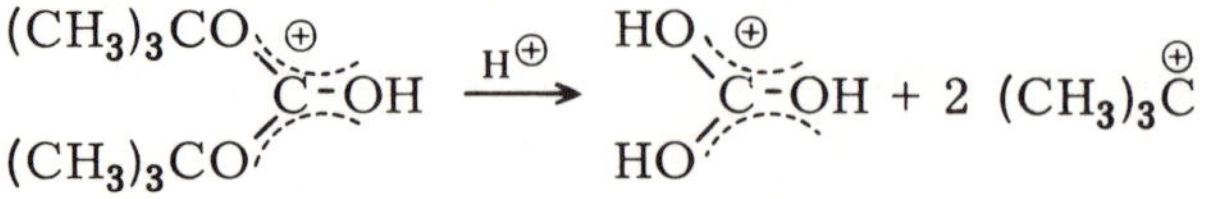

(94)

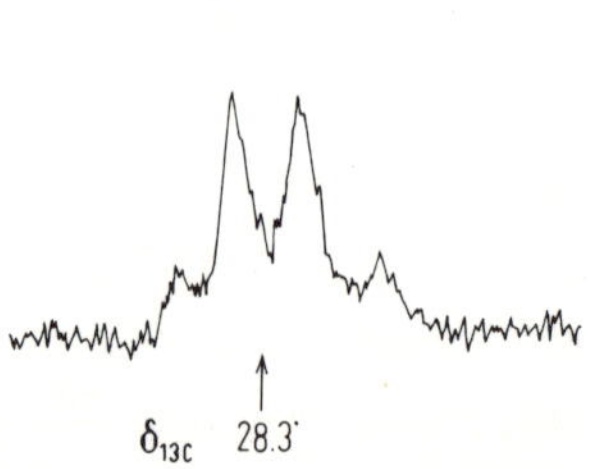


Fig. 19. $^{13}C$-INDOR spectrum of the 56% $^{13}C$-enriched protonated carbonic acid *(94)* (trihydroxycarbenium ion).

Di-isopropyl and diethyl carbonate cleave at a higher temperature, also *via* alkyl-oxygen cleavage, with initial formation of the protonated monoalkyl hydrogen carbonates *(95)*, which can also be obtained by protonation of their sodium salts. Upon further heating cleavage takes place to give protonated carbonic acid.

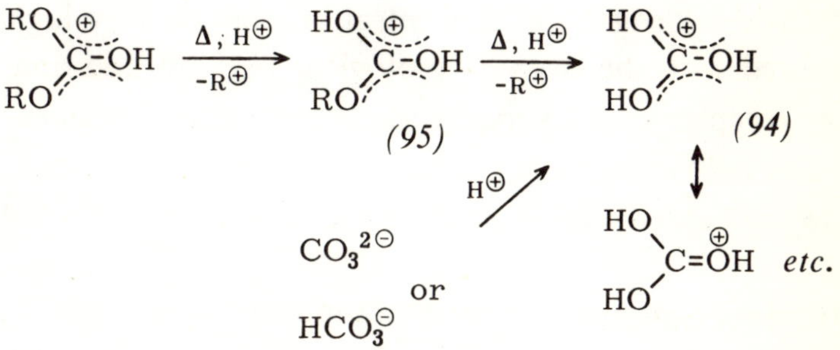

Protonated carbonic acid *(94)* can also be obtained by dissolving inorganic carbonates and hydrogen carbonates in $FSO_3H—SbF_5$ at $-80\,^{\circ}C$. It is stable in solution to about $0\,^{\circ}C$, when it decomposes to the hydronium ion and carbon dioxide. Protonated carbon dioxide ($CO_2H^+$) may be an intermediate.

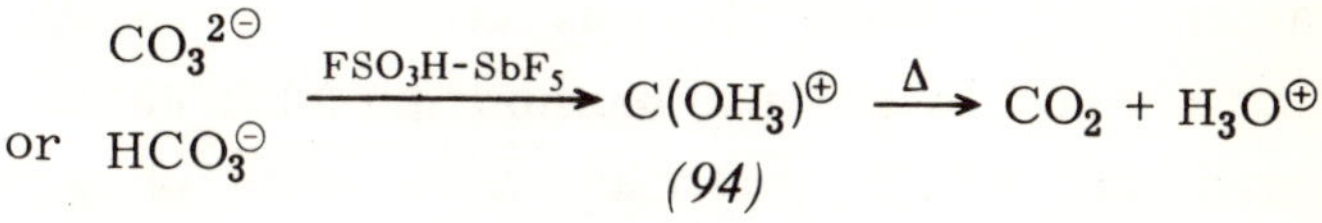

It is worthwhile to point out the close similarity between protonated carbonic acid *(94)* (the trihydroxycarbenium ion) and its nitrogen analog the guanidinium ion. Both are highly resonance stabilized through their onium forms.

The existence of protonated carbonic acid as a stable chemical entity with substantial resonance stabilization may have major implications in our understanding of some of the more fundamental biological carboxylation processes. Obviously the *in vitro* observation in specific, highly acidic, solvent systems cannot be simply extrapolated to different environments (biological systems). However, it is possible that on the active receptor sites of enzyme systems (for example, those of the carbonic anhydrase type) local hydrogen ion concentration may be very high compared with the overall "biological pH". In addition, on the receptor sites a very favorable geometric configuration may help to stabilize the active species, a factor which cannot be reproduced in model systems *in vitro*.

### 2.13.2. Acyl Cations

In 1943 *Seel*[139] characterized the first stable acyl cation. Reaction of acetyl fluoride with boron trifluoride gave a complex (decomposition point 20 °C) which was characterized as the acetyl tetrafluoroborate salt.

$$H_3C{-}COF + BF_3 \rightleftharpoons H_3C{-}CO^{\oplus}BF_4^{\ominus}$$

The identification was based on analytical data and chemical behavior. Only in the 1950's was further characterization of the complex by physical methods possible, namely by infrared and NMR spectroscopy. A series of acyl cations[140-142] (oxocarbenium ions) has been isolated and identified. The hexafluoroantimonate and hexafluoroarsenate complexes were found to be particularly stable[142]. *Deno et al.* investigated solutions of carboxylic acids in sulfuric acid and oleum[143]. They observed protonation at lower acid concentrations, and dehydration to give acyl cations at higher acidities.

$$R\text{-}CO_2H \xrightleftharpoons{H_2SO_4} R\text{-}\overset{\oplus}{C}O_2H_2 \xrightleftharpoons{\text{Oleum}} R\text{-}\overset{\oplus}{C}{=}O + H_3O^{\oplus}$$

The investigation of acyl cations like the acetyl, propionyl, butyryl, isobutyryl, pivaloyl and benzoyl cation was substantially helped by NMR studies. Not only $^1H$, but also $^2H$, $^{13}C$, and $^{19}F$ resonance studies established the structure of these ions[142, 144]. The $^{13}C$- and $^1H$-NMR spectra showed that in the case of acyl cations, such as the acetyl cation *(89)*, a resonance hybrid of the oxonium ion, oxocarbenium ion, and of the ketene-like form is involved.

$$H_3C{-}C{\equiv}O^{\oplus} \longleftrightarrow H_3C{-}\overset{\oplus}{C}{=}O \longleftrightarrow \underset{H^{\oplus}}{CH_2}{=}C{=}O$$

Recent x-ray crystallography studies of the $CH_3CO^+{-}SbF_6^-$ complex substantiated this suggestion and provided convincing evidence for the linear structure of the crystalline complex[145].

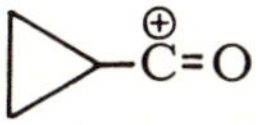
*(96)*

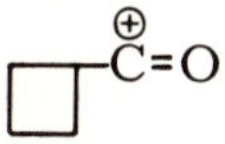
*(97)*

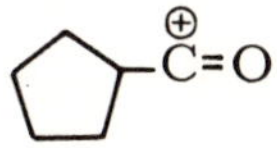
*(98)*

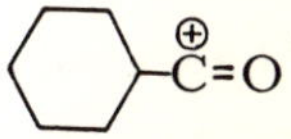
*(99)*

$O{=}\overset{\oplus}{C}(CH_2)_3\overset{\oplus}{C}{=}O$ *(100)*

$O{=}\overset{\oplus}{C}(CH_2)_4\overset{\oplus}{C}{=}O$ *(101)*

*(102)*

$H_2C{=}\underset{CH_3}{C}{-}\overset{\oplus}{C}{=}O$ *(103)*

$H_2C{=}CH{-}\overset{\oplus}{C}{=}O$ *(104)*

$H_3C{=}CH{=}\underset{CH_3}{C}{-}\overset{\oplus}{C}{=}O$ *(105)*

Investigation of acyl cations has been extended to the study of acyl cation, *i.e.* cycloacyl cations such as *(96)* to *(99)*[146], unsaturated acyl cations such as *(103)* to *(105)*[147], and diacyl cations such as *(100)—(102)*[148].

## 2.14. Sulfur as Heteroatom

Thiols and sulfides are protonated on sulfur and give, in superacid media, the stable mono- and dialkylsulfonium ions respectively[47]. Thiocarboxylic acids, *O*- and *S*-alkyl thioesters, dithioesters, and thiocarbonates also form stable thiohydroxy or thioalkoxycarbenium ions[158]. Examples are the ions *(106)—(111)*.

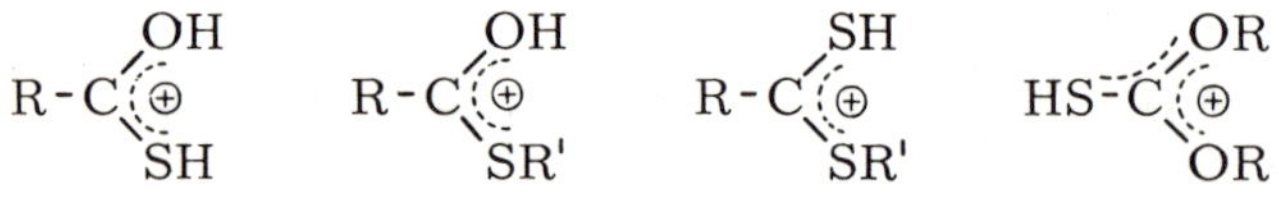

*(106)* *(107)* *(108)* *(109)*

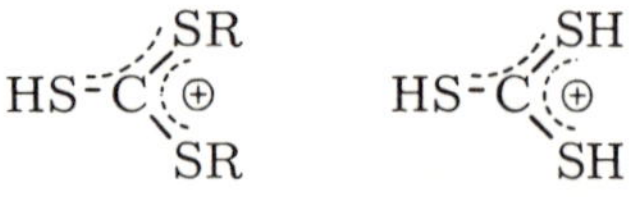

*(110)* *(111)*

## 2.15. Nitrogen as Heteroatom

Amides are protonated in superacid media at low temperatures on the carbonyl oxygen atom, as was first shown by *Gillespie*[149].

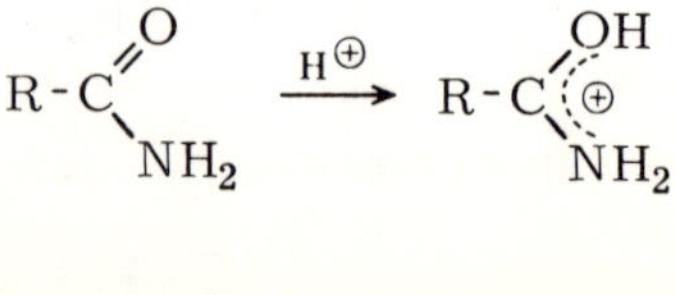

It has been claimed[150] that protonation of ethyl *N*,*N*-diisopropylcarbamate, a hindered amide, takes place on nitrogen and not on oxygen. A reinvestigation, however, established that at low temperature *O*-protonation first takes place (kinetic control) and that the *O*-protonated amide subsequently rearranges to the more stable *N*-protonated form (thermodynamic control)[151].

The possibility of observing the protonated amide linkage in strong acid media has particular relevance in the study of peptides and proteins[152, 153].

Since nitrogen is a much stronger electron donor than oxygen, the contribution of carbenium ion structures in acid salts of imines, amidines, and guanidines is small[154].

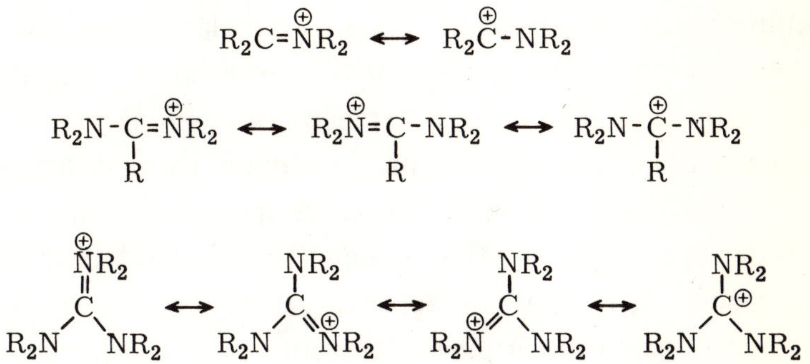

However, even in protonated nitriles contribution from the iminocarbenium ion resonance form must also be considered[155, 156].

$$H_5C_2{-}C{\equiv}\overset{\oplus}{N}H \longleftrightarrow H_5C_2{-}\overset{\oplus}{C}{=}NH$$

Ions such as the triazocarbenium ion, $(N_3)_3C^+SbCl_6^-$, have also been reported[157].

The field of protonated heteroaliphatic compounds has been studied in detail; interested readers are referred to a recent review[158] for more information.

## 3. Equilibrating Carbenium Ions or Static Carbonium Ions? The Classical-Nonclassical Ion Controversy

Some carbocations show great tendency to undergo fast degenerate rearrangements, leading through intramolecular hydrogen or alkyl shifts to the related identical structures[2]. The question arises, whether these processes are equilibrations between the limiting carbenium ("classical") ion intermediates separated by low energy level transition states or whether intermediate hydrogen or alkyl bridged carbonium ions are involved. Extensive discussions regarding the kinetic and stereochemical results in these systems have already appeared in the literature and it is considered outside the scope of this review to recapitulate the arguments. The reader is referred to reviews, such as *Schleyer*'s[2], and the original literature.

As direct observation of long-lived carbocations, generally in superacid media, became possible in recent years it was as a logical extension of our work to apply the developed spectroscopic and chemical studies of direct observation to the study of equilibrating or bridged carbocation systems. Differentiation between these two possibilities by NMR spectroscopy is difficult; because of the "slow time scale" of the NMR method, rapidly equilibrating or bridged ions can be expected to give similar $^{1}$H-NMR spectra.

$^{13}$C-NMR spectra can, however, be used advantageously[30] for investigation of the structure of carbocations in which the possibility exists for degenerate rearrangements to occur which are fast with respect to the NMR time-scale and which lead to average shifts and coupling constants being observed. "Rapid" methods such as IR, Raman, and especially ESCA spectroscopy are therefore particularly useful for investigating these systems.

## 3.1. Equilibrating Alkylcarbenium Ions

The dimethylisopropylcarbenium ion is an example of a rapidly equilibrating alkylcarbenium ion. Its $^{1}$H-NMR spectrum, shows four equivalent methyl groups and does not permit a clear distinction to be made between a rapidly equilibrating system or a static ion [which may be formulated as involving hydrogen bridging, *i.e.* a π-complex *(112c)* or alkenonium ion *(112d)*].

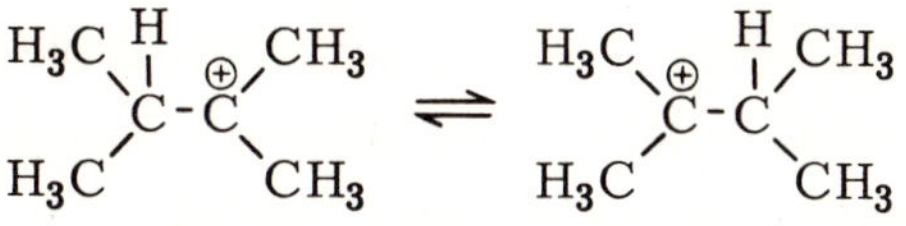

*(112a)* *(112b)*

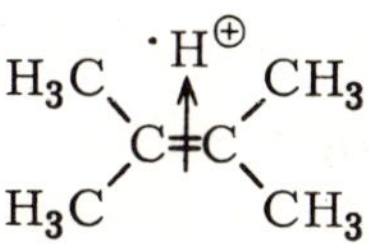

*(112c)*

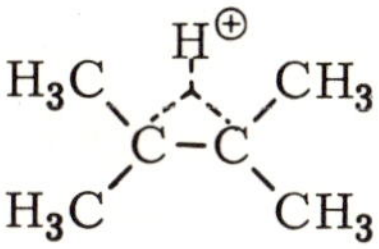

*(112d)*

The average $^{13}$C chemical shift observed for the two central carbons of the dimethylisopropylcarbenium ion is δ 3.4. The INDOR spectrum consists of a doublet

($J_{CH}$=65 Hz)[27]. In the case of an equilibrating system the average value of the shifts at the two sites, *i.e.* $C^+$ and HC<, would be observed as the chemical shift. A suitable model for estimating these shifts is the *tert*-butyl cation; the average value of the $^{13}C$ chemical shift of the methyl group and of the central $C^+$ atom is $\delta_{^{13}C}$ 6.2.

The effect of the two additional methyl groups in *(112)* should cause additional deshielding of both shifts compared with those in the model compound, and thus there is excellent agreement between the observed shift and the chemical shift expected for the rapidly equilibrating ion.

The coupling constant can also be estimated (again using the *tert*-butyl cation as a model compound) from the direct and long-range C—H coupling constants $J_{HC}$ and $J_{HCC}$. The average of these values (assuming the long-range coupling to be negative, as is usual in three-bond CCH coupling) is 64 Hz, again in excellent agreement with the observed value. The agreement of both the coupling constants and the chemical shift with the values predicted for the rapidly equilibrating ion are so good as to leave little doubt as to the nature of this ion. IR and Laser Raman spectroscopic studies of the ion also confirm this conclusion. Whereas NMR studies of rapidly equilibrating ions are thus possible, clearly the relatively slow time-scale of the NMR experiment does not allow direct, separate observation of individual species, but gives information through analysis of *average* chemical shifts and coupling constants of the equilibrating species. Thus it is clear that the study of these ions necessitates the use of physical methods whose time scale is not affected by even the fastest chemical equilibrating processes.

Consequently we carried out a study of the infrared and Raman spectra[25, 32] of the isopropyldimethylcarbenium ion and, for comparisons, those of a series of alkylcarbenium ions with known "static" structure, such as the *tert*-butyl, *tert*-amyl, and isopropyl cations. The nearly identical spectra of the ions and the evident planarity (or near planarity) of the carbenium centers suggest that the dimethylisopropylcarbenium ion is "classical", similarly to the static ions used for comparison.

Finally, the most decisive physical method yet applied to the study of carbocations in the condensed state is perhaps that of x-ray photoelectron spectroscopy (ESCA)[36], which allows one to differentiate between equilibrating carbenium ions and bridged carbonium ions. This method allows direct measurement of carbon 1 s electron binding energies. As the charge distribution within carbocations causes increasing binding energies with increasing positive charge localization, highly electron deficient classical alkyl and cycloalkyl carbenium centers (as in *tert*-butyl and *tert*-amyl cations) show at least 4 eV binding energy differences from the remaining less electropositive carbon atoms.

In the ESCA spectra of bridged carbonium ions, such as the norbornyl cation (see Section 3.3), there is no indication of such a highly electron deficient carbon center with high binding energy. The ESCA spectra of the dimethylisopropylcarbenium ion *(112)* and of the dimethylethylcarbenium ion *(113)*...

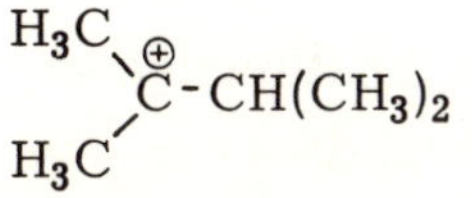


*(112)*

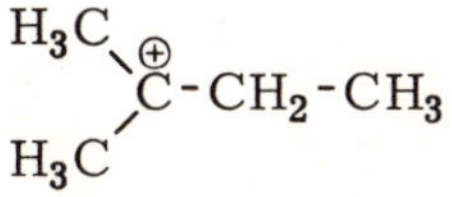


*(113)*

benium ion *(113)* show practically the same difference in the 1s bonding energies ($\Delta E_b$) between $C^+$ and adjacent C atoms ($\approx$4.2 eV).

Since ESCA spectra represent observation of the ejected core electron from a single species, no time scale limitation related to possible chemical equilibration phenomena can exist. The ESCA spectra thus clearly indicate that the ion studied is a classical carbenium ion and not a bridged tetramethylethyleneprotonium ion *(112d)*.

An example for differentiating between an equilibrating and a methyl bridged ion is the dimethyl-*tert*-butylcarbenium ion *(114)* (the triptyl cation or pentamethylethyl cation). The $^1$H-NMR spectrum of this ion consists of a single signal at δ 2.90[45]. The position of this chemical shift was considered indicative of a rapidly equilibrating structure [*(114a)*⇄*(114b)*] rather than a bridged structure *(114c)*, although the fact that only a single proton

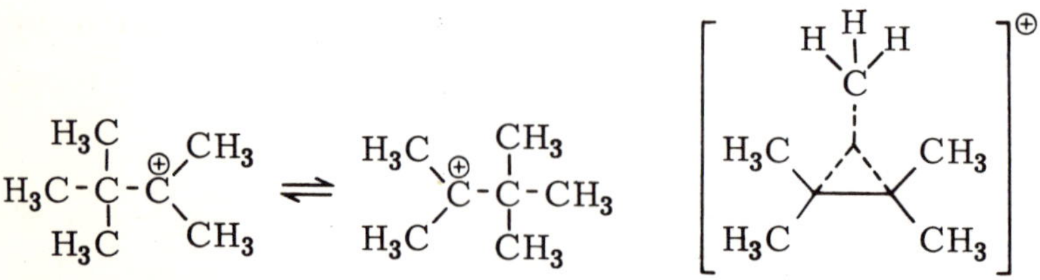


*(114a)* *(114b)* *(114c)*

resonance is observed does not necessarily enable these structures to be distinguished since in the methyl bridged carbonium ion intramolecular equilibration of the methyl groups could take place. As in the case of the dimethylisopropylcarbenium ion the $^{13}$C-NMR spectral data can give additional evidence about the structure[30].

The average $^{13}C$ chemical shift for C-2 and C-3 was found to be $\delta_{^{13}C}$ −11.5; this is consistent only with the rapidly equilibrating structure. The norbornyl cation (see Section 3.3), which has been experimentally observed, provides a good model for a methyl bridged ion. The bridged carbon atoms in this ion have a chemical shift $\delta_{^{13}C}$ 70. If the dimethyl-*tert*-butylcarbenium ion is similar in nature to the norbornyl cation we would expect the C-2 and C-3 shifts in the former compound to be 81.5 ppm further upfield than they actually are. The near zero coupling between the methyl protons and carbon-13 is also revealing. Since this ion rapidly equilibrates, the observed coupling will be the average of the two and three bond carbon-hydrogen coupling constants. The two bond coupling constant in the *tert*-butyl cation is 6.5 Hz. A similar value is to be expected for the methyl bridged ion. Since the observed coupling constant is zero, the three bond coupling must have the same magnitude but opposite sign. So far the signs of only a few such couplings have been determined. However, such a sign alternation has been reported by *Karabatsos*[31a], the two bond coupling being negative and the three bond coupling positive.

The $^{1}H$-NMR spectrum of the 2-butyl cation *(115)* at −120 °C consists of two signals at δ=3.2 and 6.7, the intensity ratio being 2:1[159]. This is the result of a degenerate 1,2 hydrogen shift, which at this temperature is fast with respect to the NMR time-scale.

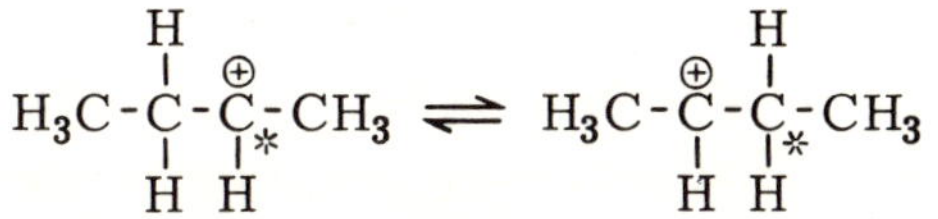


*(115)*

In the $^{13}C$-NMR spectrum the two central carbons give a quartet at $\delta_{^{13}C}$ 21.2 with a carbon-hydrogen coupling constant of $J_{CH}=70\pm2$ Hz[30]. Using the difference of the $^{13}C$ shifts observed in the dimethylisopropyl ion *(112)* and the dimethyl-*tert*-butylcarbenium ion *(114)* as a measure of the effect of methyl substitution on the $^{13}C$ chemical shift in equilibrating ions (8 ppm) then a $^{13}C$ chemical shift of $\delta_{^{13}C}$ 13 is to be expected. This is in reasonable agreement with the observed value. Using the isopropyl cation as a model gives an estimated coupling constant of 72 Hz. As in the two previous examples the $^{13}C$-NMR values clearly demonstrate that the 2-butyl cation is an equilibrating "classical" carbenium ion rather than a bridged "non-classical" carbonium ion. Both Raman and ESCA spectroscopic studies of the pentamethylethyl and 2-butyl cations clearly indicate the classical carbenium ion nature of the ions. The 1s binding energies of the carbenium centers and the remaining carbon atoms differ in the case of the tertiary ions by about 4.0—$4.8\pm0.2$ eV.

### 3.2. The Equilibrating Cyclopentyl Cation

As already mentioned the $^{1}H$-NMR spectrum of the cyclopentyl cation *(116)* in $SbF_5$—$SO_2ClF$ solution at $-70$ °C consists of a singlet at $\delta$ 4.68. The $^{13}C$ satellites of this peak show a $J_{CH}$ coupling constant of 28.5 Hz and have an intensity five times that normally observed. This results from the degenerate rearrangements that occur in this ion which, on the NMR time-scale, lead to equivalence of the nine protons and five carbon atoms. The $^{13}C$-

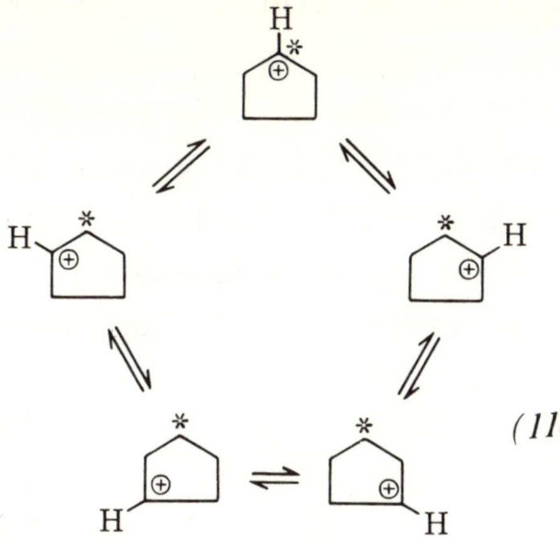


*(116)*

INDOR spectrum shows a 10 line multiplet at $\delta_{^{13}C}$ 95.4 with a coupling constant of 28.5 Hz[30] (Fig. 20).

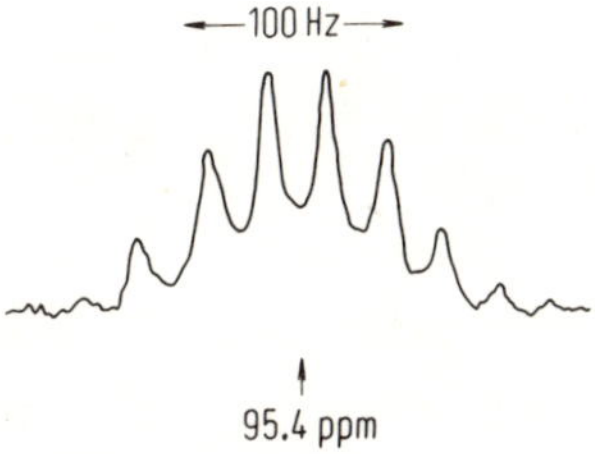


Fig. 20. 25 MHz $^{13}$C-INDOR spectrum of the cyclopentyl cation obtained in natural abundance from cyclopentyl chloride in $SO_2ClF$—$SbF_5$ solution. The ten-line multiplet ($J_{CH}$ = 28.5 Hz) is a result of the degenerate rearrangements in this ion leading to complete scrambling of the nine protons among the five carbons on the NMR time scale. The observed shift and coupling constant are thus average values.

Once again the ESCA spectrum[36], which is not affected by the time scale of equilibration, shows that the ion *(116)* is a secondary carbenium ion. The binding energies of the carbenium carbon atom and remaining carbon atoms differ by about 4.8 eV.

## 3.3. The Norbornyl Cation

The cyclopentyl cation *(116)* leads us to the discussion of one of the most controversial of all carbocations, the norbornyl cation *(117)*, whose structure has been the central point of many discussions regarding the much publicized classical-nonclassical controversy of "carbonium ions"[9].

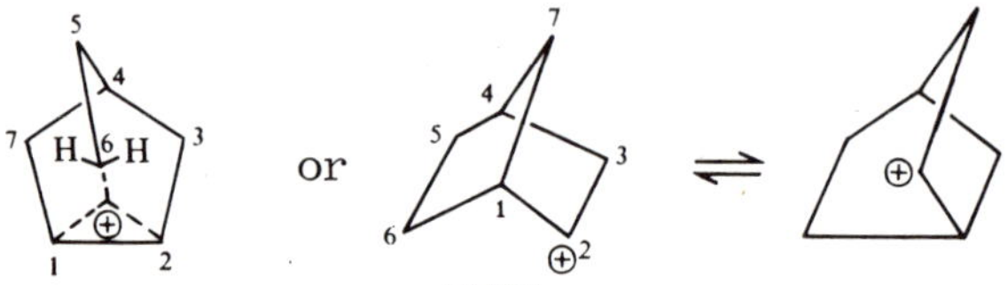


*(117)*

The methods worked out in our laboratories to generate and study stable carbocations in solution were successfully applied to the direct observation of the norbornyl cation and the clarification of its structure[10].

In 1964 *Saunders*, *Schleyer*, and *I* first investigated[160] the ion prepared from 2-*exo*-fluoronorbornane *(118)* in $SbF_5$—$SO_2$ and recorded its $^1$H-NMR spectrum. At room temperature it consists of a single broad band at $\delta = 3.75$ (Fig. 21) due to scrambling of all hydrogen atoms. The

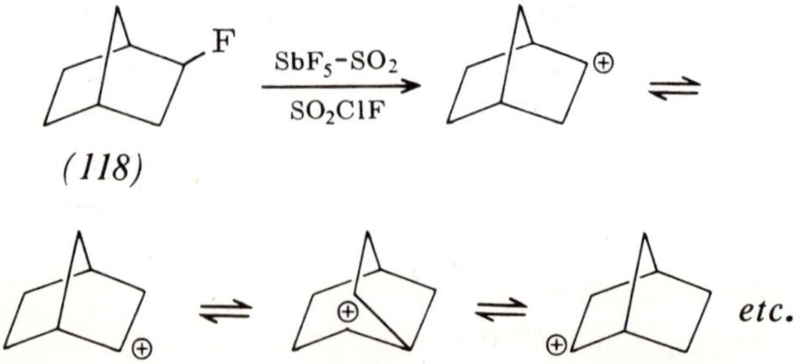


*(118)*

contrast between this single band and the complex spectrum of the starting compound is striking. The equilibration of the hydrogen atoms is caused by the fast 3,2- and 6,2-hydrogen shifts and the Wagner-Meerwein rearrangement.

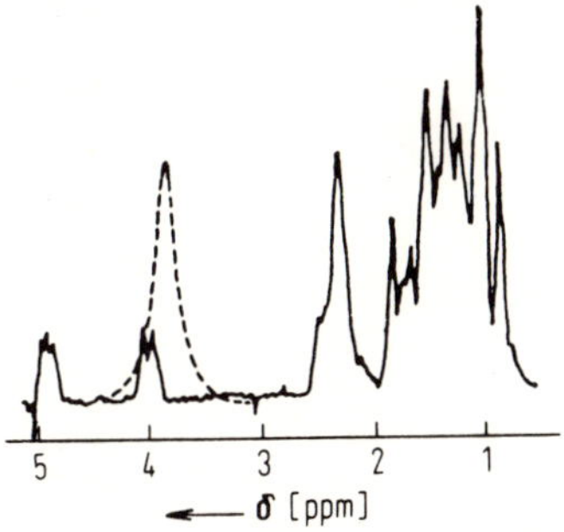


Fig. 21. 60-MHz $^{1}H$-NMR spectrum of the norbornyl cation *(117)* (– – –) and the 2-*exo*-fluoronorbornane *(118)* (——) at 35 °C.

Similarly at room temperature the $^{13}C$-NMR spectrum of the ion shows a single broad absorption line centered at $\delta_{^{13}C}$ 134[10].

When the temperature is lowered to –60 °C the $^{1}H$-NMR spectrum of the 2-norbornyl cation is resolved into three different peaks (intensity ratio 4:1:6). The spectrum does not change when the temperature is lowered to –120 °C[10] (Fig. 22).

The spectrum was interpreted as evidence that the 3,2-hydrogen shift could be frozen out, but the 6,2-hydrogen shift and the Wagner-Meerwein rearrangement are still fast even at the low temperatures used (–120 °C). The rate constant of the slow 3,2-hydrogen shift was established from temperature-dependence studies, and the activation energy of this shift was found to be 12 kcal/mol. In studies

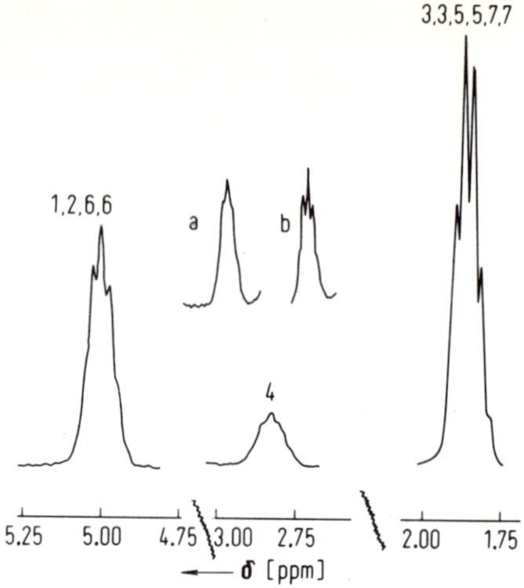


Fig. 22. $^1$H-NMR spectrum (100 MHz) of the norbornyl cation *(117)* in $SbF_5$—$SO_2$ solution at −80 °C. The insert shows the effect of irradiating the low-field septuplet on the one-proton peak at δ = 2.82. a is the spectrum without irradiation and b is with irradiation.

with *White*[10] we subsequently succeeded in "freezing out" on the NMR time scale even the fast 6,2-hydrogen shift. Using a mixed solvent system ($SbF_5$—$SO_2ClF$—$SO_2F_2$) it was possible to observe the 100 MHz spectrum at temperatures down to −156 °C. At −120 °C, the spectrum was identical to that described previously; however, in the temperature range −128 to −150 °C significant changes occur. The most deshielded peak due to the four equilibrating "protonated cyclopropane" ring protons broadens and then separates into two signals at δ 3.05 and 6.59 with relative intensities 2:2. The high field resonance, due to the six methylene protons, also broadens and develops a shoulder at δ 1.70. The signal at δ 2.82 due to the single bridgehead proton remains unchanged (Fig. 23).

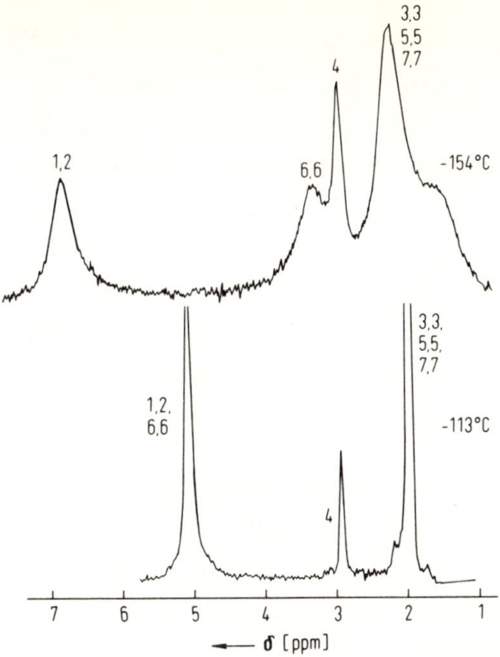


Fig. 23. 100-MHz $^{1}$H-NMR spectrum of the "norbornyl cation" in $SbF_5$—$SO_2ClF$—$SO_2F_2$ solution at temperatures between −113 and −154 °C.

The temperature dependence of the low field resonance was used to calculate the rate constants based on the Gutowsky-Holm equations. The activation energy from the Arrhenius plot is $5.9 \pm 0.2\,\mathrm{kcal/mol^{-1}}$ and the pre-exponential factor $10^{12.7}\,s^{-1}$.

These observations could signify either: (a) that the ion is "classical" and that the temperature dependence corresponds to the "freezing out" of the 6,1,2-hydrogen shift, the Wagner-Meerwein rearrangement still being fast at −156 °C, or (b) that the ion is "non-classical", all rearrangements have been "frozen out" and thus the structure is that of a methylene bridged ion with a pentacoordinated carbon atom.

In order to differentiate between these two possibilities the Raman and $^{13}$C-NMR spectra of the ion were studied[10]. As Raman spectroscopy is a rapid physical method (assuming that vibrational transition rates are faster than any of the hydrogen or alkyl shifts) the question of possible equilibration *versus* bridged ion becomes unimportant. The previously discussed technique of average $^{13}$C-NMR shifts should be also applicable in this case for differentiating a static nonclassical bridged ion from rapidly equilibrating classical carbenium ions.

The Raman spectroscopic investigation of the stable norbornyl cation was carried out at $-70\,°C$ in $FSO_3H—SbF_5—SO_2$ solution using a helium-neon laser. In the C—C stretching frequency region there is only one strong line at 972 cm$^{-1}$ ($\rho = 0.35$). The C—H stretching frequency region shows five lines, one of which is at 3110 cm$^{-1}$. The skeletal vibrations of the ion and their comparison with those of model compounds suggest a structure closer related to nortricyclene, but not to norbornane, *i.e.* under the conditions of the investigations ($-70\,°C$, where, as the NMR spectra show, the ion rapidly equilibrates) an ion having protonated nortricyclene nature is present.

As mentioned, the $^{13}$C-NMR spectrum of the ion at room temperature shows a single broad absorption. On lowering the temperature to $-70\,°C$, three well resolved absorptions are observed (Fig. 24). The chemical shift of the triplet due to the equivalent methylene carbon atoms C-3, C-5 and C-7 is at $\delta_{^{13}C}$ 162.5 ($J = 140.2$ Hz) and that of the bridgehead carbon atom C-4 at $\delta_{^{13}C}$ 156.1 (doublet $J = 153$ Hz). The 53.3 Hz pentuplet observed at $\delta_{^{13}C}$ 101.8

(from $^{13}CS_2$) using the INDOR method (on irradiation of the satellites of the lowest field peak) corresponds to the three equivalent carbons C-1, C-2 and C-6 associated with the four equivalent equilibrating protons.

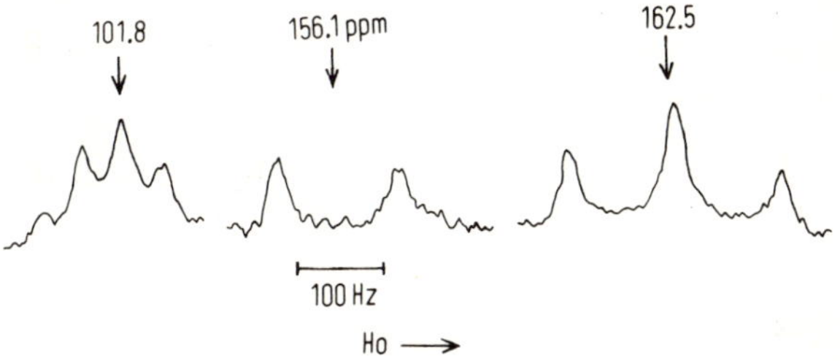


Fig. 24. $^{13}$C-INDOR spectrum of the norbornyl cation *(117)* generated in $SbF_5$—$SO_2$ solution from *exo*-2-chloronorbornane and recorded at −70 °C. The spectrum was obtained by observing the $^{13}$C satellites in the 100-MHz $^1$H-NMR spectrum while irradiating the sample with a swept 25-MHz frequency. $^{13}$C shifts indicated are with respect to $^{13}CS_2$.

Subsequently we were also able to obtain the $^{13}$C-NMR spectrum of the "frozen out" norbornyl cation at −156 °C. Whereas the use of the INDOR method is generally based on observing the effects of irradiation on the $^{13}$C satellites in the proton spectrum, these satellites could not be observed at −156 °C due to a poor signal-to-noise ratio at this very low temperature. Enhancement of the C-1, C-2-proton resonance (due to a collapse of the $^{13}$C satellites and possibly a nuclear Overhauser effect) proved to be possible on irradiation at 25.1 MHz, however, even at −156 °C. The low-field proton resonance (at δ 6.59) was enhanced on irradiation at $\delta_{^{13}C}$ 70 ± 2 while the more shielded proton resonance (at δ 3.05) was enhanced on irradiation at a frequency corresponding to $\delta_{^{13}C}$ 173 ± 2.

We were unable to obtain the C—H coupling constants and peak multiplicities in this manner. Recently, however, it has been possible by use of the rapid Fourier Transformation Method to obtain the complete $^{13}$C-NMR spectrum, and all the coupling constants and peak multiplicities. Table 3 lists the corresponding values of the carbon atoms participating in the three center bond.

It should also be mentioned that both the "σ-route" from 2-norbornyl halides and the "π-route" from β-$\Delta^3$-cyclopentenylethyl halides lead to the same norbornyl cation *(117)*.

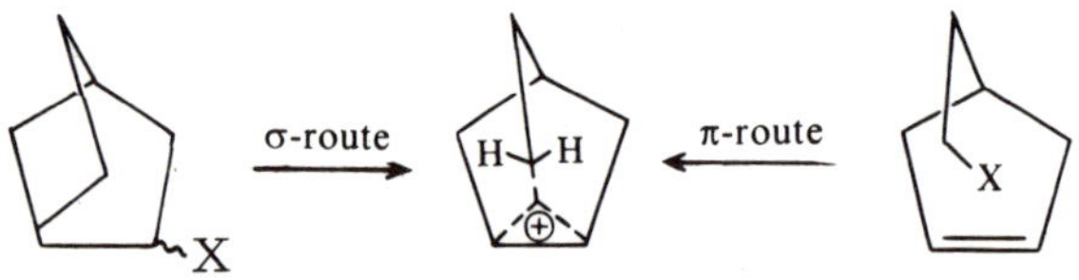


*(117)*

Recently *Mateescu, Riemenschneider*[36], and *I* succeeded in recording the ESCA spectrum of the norbornyl cation and compared it with that of the 2-methylnorbornyl cation and other trivalent carbenium ions, such as the cyclopentyl and methylcyclopentyl cations *(116)* and *(37)*. The 1s electron spectrum (Fig. 25) of the norbornyl cation *(117)* shows no high binding energy carbenium center and a maximum separation of less then 1.5 eV between the two "cyclopropyl" type carbons, to which bridging takes place, from the other carbon atoms (including the pentacoordinated bridging carbon). In contrast the 2-methylnorbornyl cation shows, like other trivalent carbenium ions, a high binding energy carbenium center that is only slightly delocalized, as indicated by the binding energy difference of 3.7 eV from the other carbon atoms.

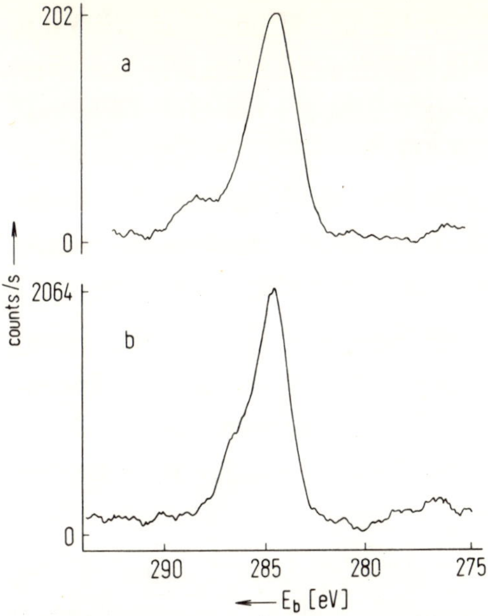


Fig. 25. Carbon 1 s electron spectrum (ESCA spectrum) a) of the 2-methylnorbornyl cation and b) of the norbornyl cation *(117)*. $E_b$ = binding energy.

Since in electron spectroscopy the time scale of the measured ionization processes is of the order of $10^{-16}$ s, definite ionic species can be characterized, regardless of possible intra- and intermolecular interactions (*e.g.* Wagner-Meerwein rearrangements, hydrogen shifts, proton exchange, *etc.*), which have no effect. Thus, ESCA spectroscopy gives an undisputable, direct answer to the long debated question of the "nonclassical" nature of the norbornyl cation independent of any possible equilibration process.

The bridging methylene carbon atom C-6 of the norbornyl ion is pentacoordinated. It is bound to two hydrogen atoms and to the carbon atom C-5 by three two-electron single bonds. The remaining $sp^3$ orbital is involved in two-electron three-center bonding with C-1 and C-2. Thus the pentacoordinated carbonium ion center is similar to $CH_5^+$ and the norbornyl ion is a carbonium ion.

Other examples of directly observed long-lived carbonium ions in which pentacoordinated carbonium centers were identified by $^{13}C$ and $^{1}H$-NMR spectroscopy are the 7-norbornenyl *(119)* and the 7-norbornadienyl cation *(120)*[10,161].

*Winstein*[162] and *Richey*[163] have shown that the 7-norbornenyl cation *(119)* can be obtained *via* both the "σ-route" and the "π-route".

The carbonium centers in the norbornyl cation *(117)*, the 7-norbornenyl cation *(119)*, and the 7-norbornadienyl cation *(120)*[10,164] are similar and related to the ethylenemethonium ion *(121)*, although strain differences in the ions are obviously significant. All these ions contain a two electron three-center bond. The bridging carbon atoms are pentacoordinated, whereas the carbons to which they bridge are tetracoordinated carbonium atoms.

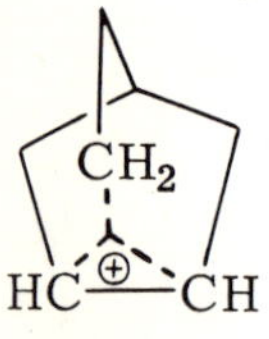


*(117)*

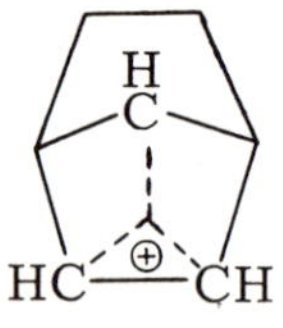


*(119)*

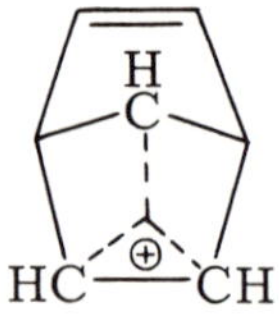


*(120)*

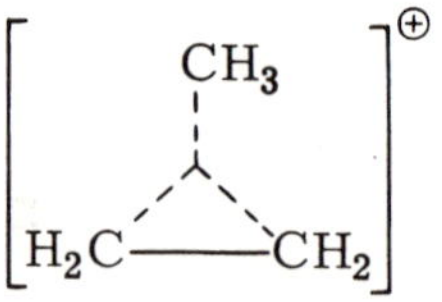


*(121)*

Comparison of the characteristic $^{13}C$-NMR parameters of the carbonium centers in these ions with those of trivalent carbenium centers in related tertiary classical ions (Table

3) shows the very significant differences which allow clear differentiation. The carbenium centers are highly electron deficient and therefore deshielded, whereas the carbonium centers show much higher shielding (up to 300 ppm) on account of their higher coordination.

It can be concluded that the original views of *Winstein*[165] on the nonclassical nature of the norbornyl cation, based on kinetic and stereochemical results, have been fully substantiated by direct spectroscopic studies of the long-lived ion[10]. Whether the ion is completely formed under solvolytic conditions ($S_N1$ type) or the reaction has more $S_N2$ character, cannot be decided at the present time. Investigation of the norbornyl cation, however, contributed substantially to our general understanding of carbonium ions as being distinct and well differentiated species from carbenium ions. Carbonium ions cannot only be formed intramolecularly, but also in intermolecular reactions and represent the key for electrophilic reactions on single bonds (see following discussion).

## 3.4. Cyclopropylmethyl and Cyclobutyl Cations

The solvolytic reactions of the cyclopropylmethyl—cyclobutyl—allylmethyl system has been studied extensively by *Roberts*[166], *Wiberg*[167], and other authors. In these studies and in recent spectroscopic investigations of the long-lived ions in superacids[76], the nonclassical nature of the common ions could be established. Rapid equilibration between the so-called unsymmetrical bicyclobutonium ions is in agreement with all data.

Table 3. Comparison of the $^{13}C$-NMR parameters of the carbocation centers in carbonium and carbenium ions.

| Ion | Carbonium ions δ[ppm] | $J_{CH}$[Hz] | $\delta_{^{13}C}$ [ppm] |
|---|---|---|---|
| 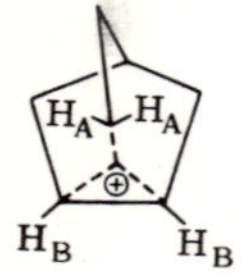 *(117)* | $H_A = 3.05$<br>$H_B = 6.59$ | $H_A = 145.8$<br>$H_B = 184.5$ | $C_A = 171.4$<br>$C_B = 68.5$ |
| 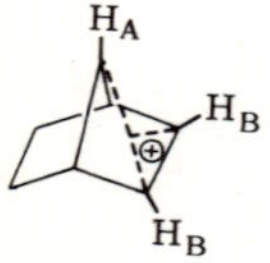 *(119)* | $H_A = 3.25$<br>$H_B = 7.04$ | $H_A = 218.9$<br>$H_B = 193.8$ | $C_A = 159.8$<br>$C_B = 67.9$ |
| 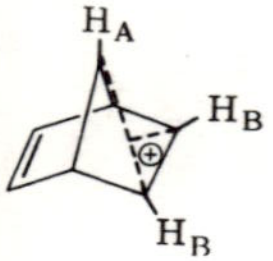 *(120)* | $H_A = 3.24$<br>$H_B = 7.48$ | $H_A = 216.4$<br>$H_B = 192.3$ | $C_A = 157.6$<br>$C_B = 78.9$ |

| Ion | | Carbenium ions Solvent $T$[°C] | $\delta_{^{13}C}$ [ppm] |
|---|---|---|---|
| 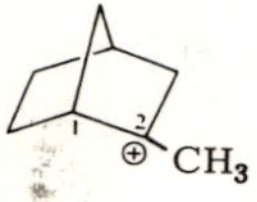 | | $FSO_3H—SbF_5—SO_2$ −80 | C-1 = +113.0 ($J$ = 169.5 Hz) C⊕-2 = −77.3 |
|  | *(37)* | $SbF_5—SO_2$ −60 | C⊕-1 = −142.0 |
| $(CH_3)_3C^{\oplus}$ | *(2)* | $SbF_5—SO_2ClF$ −60 | C⊕ = −135.4 |
| $(CH_3)_2CH^{\oplus}$ | *(1)* | $SbF_5—SO_2ClF$ −60 | C⊕ = −125.0 |
| 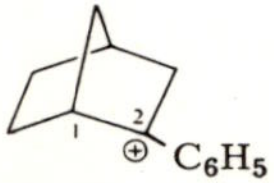 | | $FSO_3H$ −35 | C-1 = +134.0 ($J$ = 158.4 Hz) C⊕-2 = −63.5 |
| 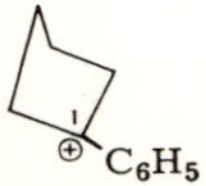 | | $FSO_3H$ −20 | C⊕-1 = −74.6 |
| $(CH_3)_2\overset{\oplus}{C}C_6H_5$ | *(42)* | $FSO_3H—SbF_5—SO_2$ −60 | C⊕ = −61.1 |

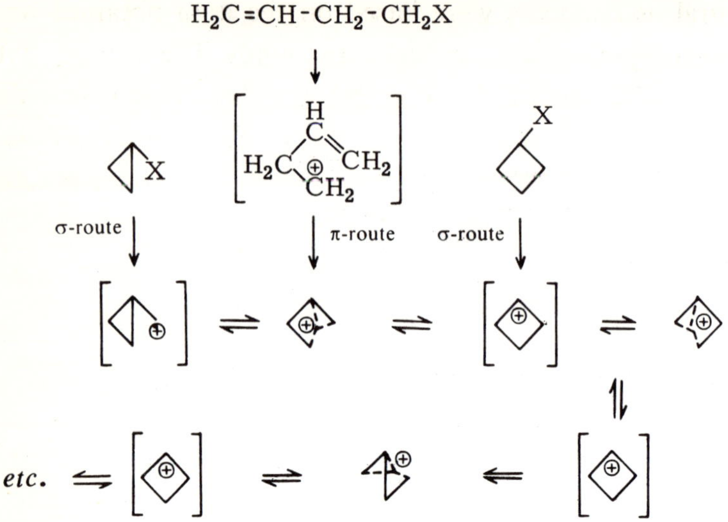


Stable, long-lived cyclopropylmethyl and cyclobutyl cations were studied in $SbF_5$—$SO_2ClF$ solution at low temperature by $^1H$- and $^{13}C$-NMR spectroscopy. Struc-

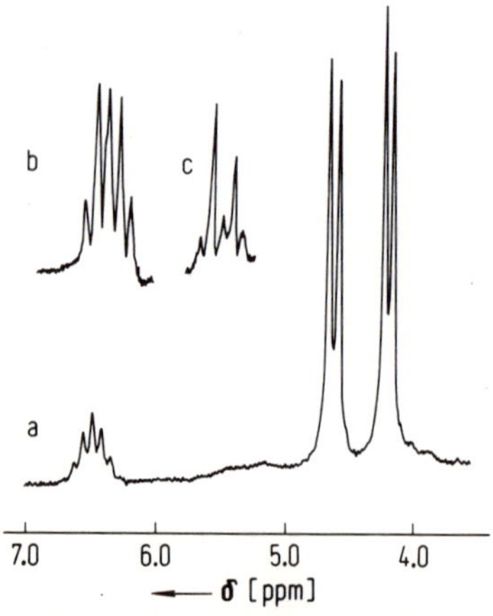


Fig. 26. a) 100-MHz $^1H$-NMR spectrum of the cyclopropylmethyl cation *(122)* in $SbF_5$—$SO_2ClF$ solution at $-80\,°C$. b) 60-MHz spectrum in the region of the CH signal; c) 60-MHz spectrum the CH signal of the α,α-dideuteriocyclopropylmethyl cation.

tural assignments were based on the data obtained and on comparison with model compounds. The cyclopropylmethyl cation *(122)* (Fig. 26) and for comparison the methyl substituted cations *(123)* and *(14)* were examined under comparable conditions. The NMR parameters are summarized in Table 4.

Table 4. NMR parameters of cyclopropylmethyl cations.

| Ion | $^1$H-NMR, δ [ppm] ($J_{H-H}$ [Hz]) | | | |
|---|---|---|---|---|
| | $CH_3$ | $CH_2$ | CH | $^{\oplus}$CH |
|  *(122)* | | 4.21 (6.5)<br>4.64 (8.0) | 6.50 (6.5, 8.0) | |
| 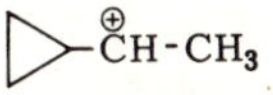 *(123)* | 3.34 (6.2) | 4.32<br>4.45 | 4.58 | 9.6 (6.2) |
| 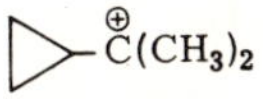 | 2.70 | 3.57 | 3.83 | |
| *(14)* [a] | 3.18 (1.25) | 3.68 | | |

| Ion | $^{13}$C-NMR, δ [ppm] ($J_{^{13}CH}$ [Hz]) | | | |
|---|---|---|---|---|
| | $CH_3$ | $CH_2$ | CH | $C^{\oplus}$ |
| *(122)* | | +138 (180) | + 85 | |
| *(123)* | +160 | +136 | +126 | −59.1 |
| *(14)* | +154 (125)<br>+163 (125) | +140 | +134 | −86.8 |

[a] See also Fig. 7.

All of the cyclopropylmethyl cations show charge delocalization by the cyclopropyl ring, but only with the primary system is the unique behavior observed which puts it in a class distinct from the secondary and tertiary ions. Comparison with model compounds, including the previously discussed norbornyl type cations, established the nonclassical carbonium ion nature of the cyclopropylmethyl cation *(122)*. A set of equilibrating cyclopropylcarbonium ions is involved, with equilibration occurring *via* the puckered cyclobutenium ion (cyclobutyl cation).

### 3.5. Ethylenearenium Ions

The classical-nonclassical ion controversy also included the question of the so-called "ethylenephenonium" ions. *Cram*'s extensive studies[168], based on kinetic and stereochemical evidence, clearly established the bridged-ion nature of β-phenylethyl cations in solvolytic systems. Spectroscopic studies (particularly $^{1}H$- and $^{13}C$-NMR)[169] of a series of stable long-lived ions proved the symmetrically bridged structure, and at the same time showed that these ions do *not* contain a pentacoordinated carbonium ion center (*i.e.* they are not "nonclassical ions"). They are spiro[2.5]octadienyl cations *(124)* (spirocyclopropylbenzenium ions) or, in other words, cyclopropylmethyl cations in which the carbenium center belongs to a cyclohexadienyl cation (benzenium ion).

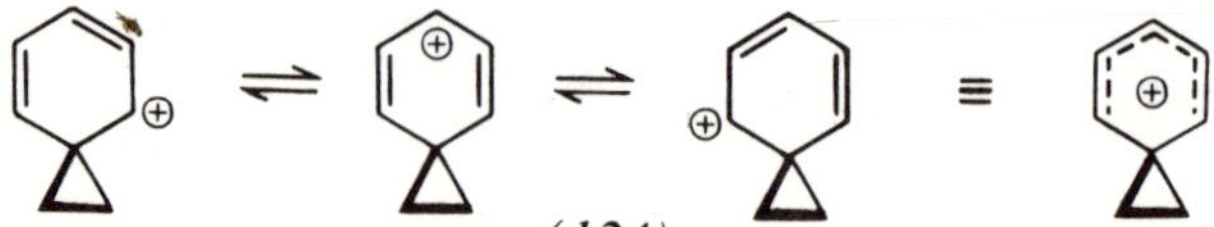

*(124)*

The nature of the spiro carbon atom is of particular importance in determining the structure of these carbocations.

$^{13}C$-NMR spectroscopic studies clearly established the aliphatic tetrahedral nature of this carbon, thus ruling out a "nonclassical" ethylenephenonium ion.

The formation of the ethylenebenzenium ion *(124)* from β-phenylethyl precursors can be depicted as cycloalkylation of the aromatic π-system and not, as such, of the $C_{Ar}$—$C_{\alpha}$ bond which would give the tetracoordinated ethylenephenonium ion. Rearrangement of the β-phenylethyl to α-phenylethyl cations on the other hand is a regular 1,2-hydrogen shift involving the $C_{\alpha}$—H bond.

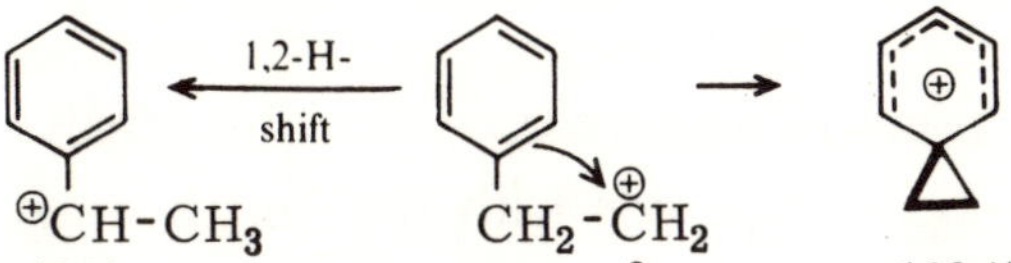


*(55)* *(124)*

# 4. The Role of Carbocations in Electrophilic Reactions

## 4.1. General Considerations

Trivalent carbenium ions, as recognized in the pioneering work of *Meerwein*, *Ingold*, and *Whitmore*, play an important role in acid-catalyzed rearrangements of hydrocarbons (isomerization, alkylation, cyclization, polymerization, *etc.*), as well in a large variety of electrophilic reactions including those of the generalized Friedel-Crafts type. These reactions and their mechanisms are well reviewed[170] and will not be discussed here. It is suggested, however, (see subsequent discussion) that only ionization of lone (non-bonding) electron pair donor precursors (n-bases) or protonation of singlet carbenes can lead directly to carbenium ions. Other (π- or σ-) donors always interact first *via* their bonding electron pairs with the electrophilic reagents, forming two electron three-center bound carbonium ions (transition states or intermediates) which subsequently yield the trivalent carbenium ions.

## 4.2. Formation of Methonium Ion (the Simplest Carbonium Ion) from Methane

The bonding of a carbonium center is considered to involve three covalent, two-electron bonds with the fourth bond being a two-electron three-center bond[12]. This type of bond occurs, similarly as in the norbornyl cation *(177)*[10], in alkonium ions like $CH_5^+$ *(125)*. The existence of the methonium ion $CH_5^+$ is based not only on mass spectrometric studies[171] but also on the chemistry of methane in superacids and with strong electrophiles[172]. Thus, the interaction involves the main lobes of the covalent bonds (frontal attack). Since an electrophile will attack the points of highest electron density, attack will occur on the covalent bonds themselves and not on the relatively unimportant back lobes.

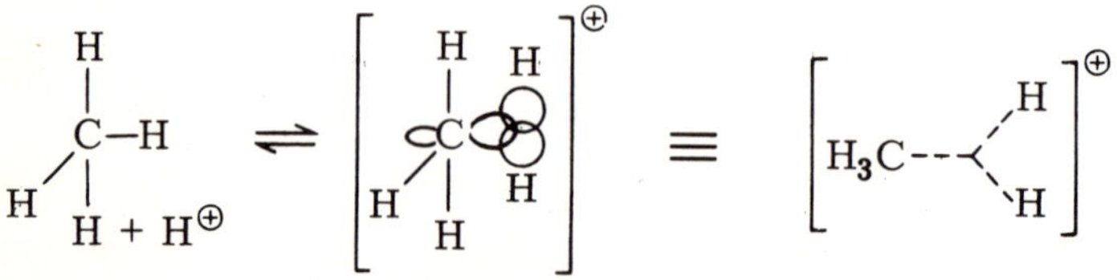

*(125)*

It should be remembered that the dotted lines symbolize the bonding orbitals of the three centered bonds and that their point of junction does not represent an additional atom.

Of the possible structures for the methonium ion ($D_{3h}$, $C_{4v}$, $C_s$, $D_{2h}$ or $C_{3v}$ symmetry) *Olah*, *Klopman*, and *Schlosberg*[173] suggested preference for the $C_s$ front-side protonated form. Preference for this form was based on consideration of the observed chemistry of methane in superacids (hydrogen-deuterium exchange and, more significantly, polycondensation indicating ease of cleavage to $CH_3^+$ and $H_2$) and also on the basis of self-consistent field (SCF) calculations[173]. More extensive calculations[174], including *ab initio* calculations utilizing an "all-geometry" parameter search[175], confirmed the favored structure with $C_s$ symmetry. This structure is about 2 kcal/mol energy-poorer than the structure with $C_{4v}$ (or $D_{2h}$ or $C_{3v}$) symmetry, which in turn is about 8 kcal/mol energy-poorer than the structure with trigonal bipyramidal $D_{3h}$ symmetry.

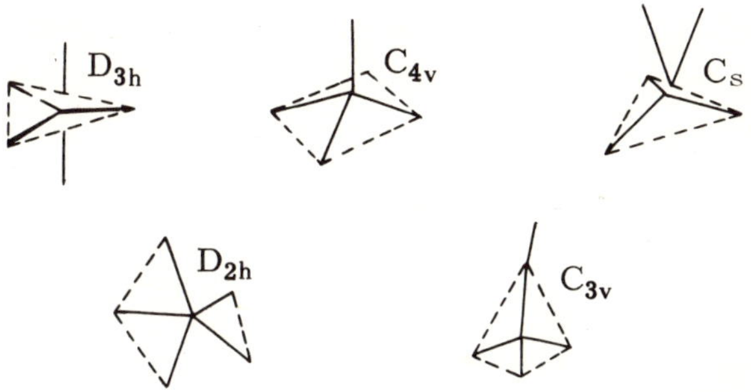


At the same time it should also be recognized that ready interconversion of stereoisomeric forms of $CH_5^+$ is possible by a pseudorotational-like process. *Muetterties* recently suggested naming stereoisomerization processes of this type in pentacoordinated compounds as "polytopal rearrangements"[176] (sometimes called "polyhedral rearrangements"). We, however, prefer to call intramolecular carbonium ion rearrangements as "bond to bond rearrangements" since these are not limited to equivalent bonds in the case of higher homologs of $CH_5^+$ (see subsequent discussion). Hydrogen-deuterium scrambling observed in superacid solutions of deuterated alkanes strongly indicate such processes.

## 4.3. Boron Analogs of Carbocations

It is of interest to note that isoelectronic boron compounds can be used as model compounds for both carbenium and carbonium ions. Trimethylboron has already been compared with the trimethylcarbenium ion (see Table 2) and it has recently been suggested that $BH_5$ may similarly be used as a model for comparison with $CH_5^+$[12].

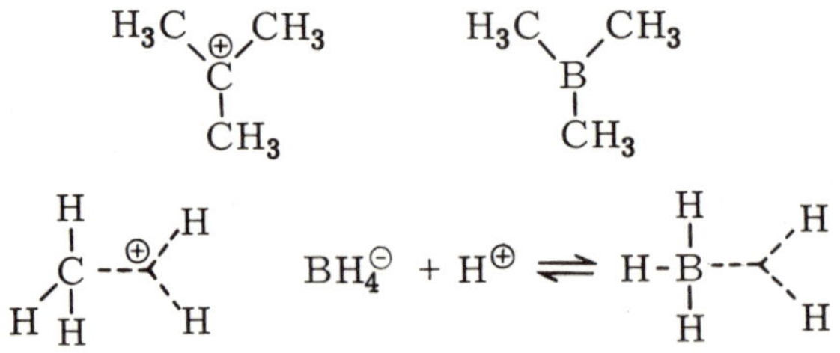

$BH_5$ is formed in the acid hydrolysis of tetrahydroborates[177,178]. When the hydrolysis is carried out with deuterated acid, not only HD but also $H_2$ is formed. In strong acids the isotopic distribution is statistical. This is indicative of $D^+$ attacking the B—H bond (similarly to the attack of $H^+$ on methane) and of polytopal (bond to bond) rearrangement of $BH_4D$ taking place, before subsequent cleavage occurs. The ease with which diborane exchanges hydrogen for deuterium when treated with deuterium gas[179] also indicates the formation of the three center bound analog, $BH_3D_2$.

$$B_2H_6 \rightleftharpoons 2\ BH_3$$

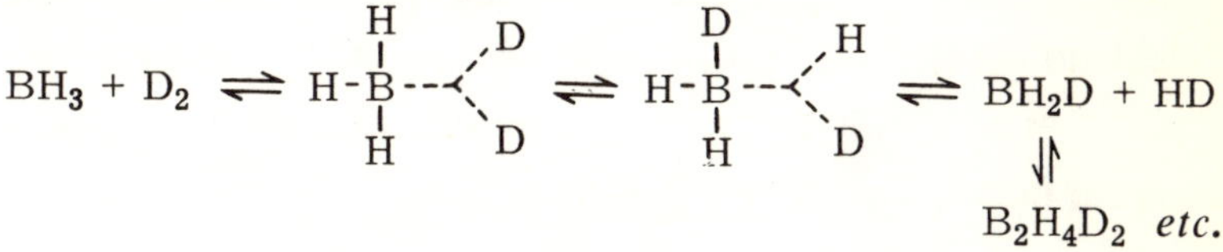


## 4.4. Steric Effects in the Formation of Carbonium Ions

It should be emphasized that if steric interference (as is the case in interaction of tertiary C—H bonds in isoalkanes with tertiary carbonium ions) is substantial, the three-center bond of carbonium ions must be highly unsymmetrical, although the attack of the tertiary carbenium ion on the tertiary hydrogen of the isoalkane still cannot be considered to lie along an extension of the C—H bond (*i.e.* in a linear fashion)[180].

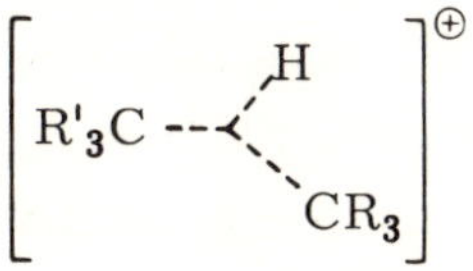


Similarly the protonation of sterically hindered C—C (and C—H) bonds can also only proceed *via* highly unsymmetrical carbonium ion states (see following discussion).

The preferred direction of attack by the electrophile should thus not be considered to be always the same and may very well vary from compound to compound (according to the reaction conditions). Since the possible configurations of pentacoordinated carbon do not differ very much in stability (as is also the case in $CH_5^+$), with exception of the "back-side" substituted trigonal bipyramid, which must be considered the least favorable form, it is unlikely that there is any "inherently" preferred pathway: it will depend on the individual reaction conditions; the electrophilic attack, however, always takes place on the appropriate bonds *via* "frontal" interaction.

## 4.5. Electrophilic Reactions at Single Bonds: σ-Donor Systems

Pentacoordinated carbonium ions play an important role not only in connection with the structure of nonclassical ions, but more importantly in the general area of electrophilic reactions at single bonds, *e.g.* in the aliphatic substitution of alkanes and cycloalkanes. Such reactions include acid-catalyzed hydrocarbon isomerizations, fragmentations, cyclizations, and other transformation reactions, electrophilic substitutions of alkanes and cycloalkanes and related electrophilic reactions.

We have in recent years obtained evidence for the general reactivity of covalent C—H and C—C single bonds of alkanes and cycloalkanes in various electrophilic substitution reactions such as protolytic processes, hydrogen-deuterium exchange[181], alkylation[180], nitration[182], halogenation[183] *etc.* This reactivity is due to what we consider, alongside n- and π-donor ability, the third major type of electron donor ability, *i.e.* the σ-donor ability (sigma basicity) of *bonded* electron pairs (*i.e.* single bonds) to share their electron pairs with an electron deficient reagent *via* two-electron, three-center bond formation. The reactivity of single bonds thus stems from their ability to participate in the formation of pentacoordinated carbonium ions. Subsequent cleavage of the three-center bond results in formation of trivalent carbenium ions and substitution products, or hydrogen (alkyl) transfer takes place.

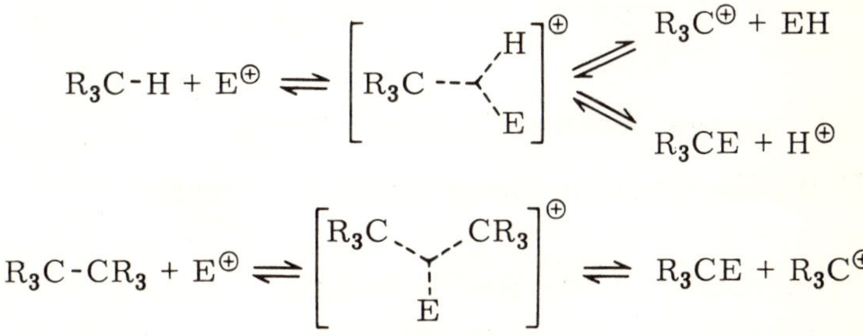

$$E^{\oplus} = D^{\oplus},\ H^{\oplus},\ R^{\oplus},\ NO_2^{\oplus},\ Hal^{\oplus}$$

As to the nature of the transition state of the electrophilic reactions at single bonds, the experimental product compositions strongly favor pentacoordinated carbonium ion-type transition states involving carbocation centers with three regular covalent two-electron bonds and one two-electron three-center bond, which, in contrast with the linear or "back-side" attack, is formed *via* "frontal" attack[12, 180].

Frontal attack involving a trigonal transition state would lead to retention of optical activity if reactions were carried out at an optically active carbon center. Experiments presently being carried out in our laboratories appear, in part, to prove this, but there are many experimental difficulties in carrying out such reactions under controlled conditions. We have also been able to carry out electrophilic substitution reactions (deuteration, alkylation, nitration, chlorination) at bridgehead positions of rigid systems, such as adamantane, where back-side attack and olefin formation is impossible (for back-side attack the electrophile would need to pass through the cage structure). These reactions give direct experimental evidence for the frontal direction of the electrophilic attack and the nature of the pentacoordinated carbonium ion-type transition state[181, 182, 184].

## 4.6. Protolysis and Hydrogen-Deuterium Exchange

With superacids, alkanes readily undergo protolytic reactions involving tertiary, secondary, and primary C—H as well as C—C bonds[172, 181]. In the case of isoalkanes, tertiary C—H bond reactivity exceeds that of C—C bonds and secondary (and primary) C—H bonds. In *n*-alkanes, the C—C bond reactivity is generally found to exceed that of the C—H bonds. It is also apparent that steric factors affect σ-basicity as they do π- or n-basicity.

*Bartlett*, *Condon*, and *Schneider*[190] were the first to recognize that in aluminum halide catalyzed intermolecular hydrogen abstractions, the electrophile (carbenium ion or proton) can remove a tertiary hydrogen atom together with its bonding electron pair (the so-called hydride transfer reaction).

$$R_3C{-}H + R^{\oplus}\ (\text{or } H^{\oplus}) \rightleftharpoons R_3C^{\oplus} + H{-}R\ (\text{or } H_2)\ .$$

In subsequent literature the assumption has generally been made that the hydrogen atom is removed with its electron pair through what amounts to a linear transition state.

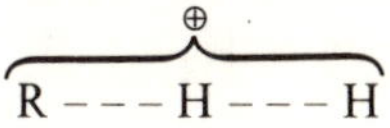

*Lewis*, *Hawthorne*, and *Symons*[185] must be given credit for having first suggested that the transition state in hydrogen abstraction could be considered triangular instead of linear: "...electrophilic reagent attacks the C—H bond and the reasonable mode of attack is on the electrons of the bond. A triangular transition state is therefore proposed." This suggestion, however, went relatively unnoticed, mainly because there were no experimental data available to substantiate it and because it was considered that steric hindrance, particularly in the interaction of a tertiary carbenium ion with a tertiary isoalkane, would allow only linear interaction.

The question of the mechanism of the hydride abstraction was not further considered until 1967, when, with *Lukas* we reported on the protolytic ionization and hydrogen-deuterium exchange of alkanes and cycloalkanes in superacids such as $FSO_3H—SbF_5(SO_2ClF)$[45,172a]. *Hogeveen*, in independent work, also made similar observations in $HF—SbF_5$ solutions[172c].

These studies demonstrated that in the new superacid systems not only tertiary and secondary, but also primary C—H bonds as well as C—C bonds undergo facile protolytic cleavage. To account for the observed protolytic reactions of alkanes, we suggested with *Klopman* and *Schlosberg*[173] that protolyses preferentially involve "front-side" attack through formation of triangular carbonium ions as shown in the case of protonation of methane to $CH_5^+$.

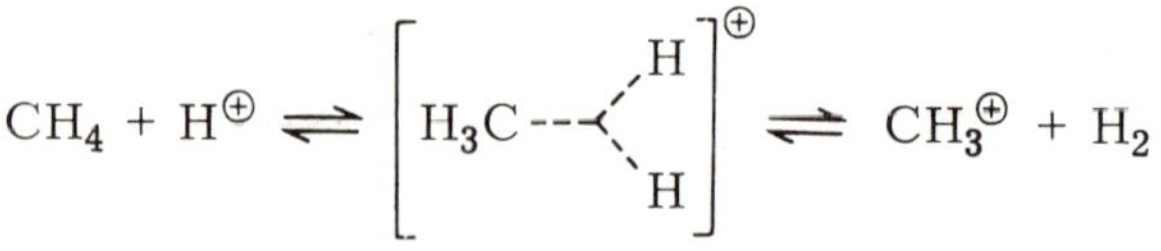

It appears, therefore, more reasonable to call these processes hydrogen abstractions (transfer) and not hydride abstractions, since *de facto* hydride ion is not involved in these processes.

Subsequently, we were able to show with *Shen* and *Schlosberg*[186] that molecular hydrogen (deuterium) itself readily undergoes hydrogen-deuterium exchange in superacids. The reaction takes place even at room temperature, and we assume that a triangular $H_3^+$-transition state *(126)* is involved in the protolytic process.

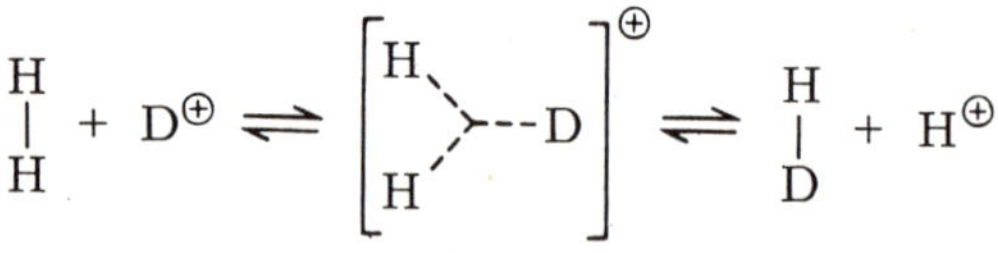

*(126)*

By way of contrast, *Hogeveen*[187a] initially rejected triangular transition states in hydride abstractions by arguing that such transition states would suffer considerable strain. Subsequently however, in collaboration with *Brouwer*, he favored bridged, unsymmetrical transition states[187b]. He also stated that he found no evidence for exchange of molecular hydrogen (deuterium) with the superacid system used (generally $SbF_5$ in a large excess of HF, which is a weaker acid system than the 1:1 acids used in our work), which he considered as further evidence against triangular transition states. At the same time he described the reaction of carbenium ions with molecular hydrogen[188]. He could not distinguish the nature of the transition state (linear *vs.* triangular) but considered the reaction to involve a linear transition state since no hydrogen-deuterium exchange was observed with molecular hydrogen (deuterium) itself.

$$R^{\oplus} + H{-}H \rightleftharpoons (R{-}{-}{-}H{-}{-}{-}H)^{\oplus} \rightleftharpoons R{-}H + H^{\oplus}$$

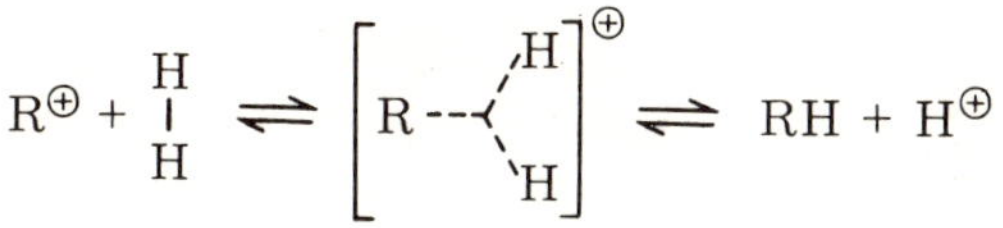

According to our studies of the hydrogen-deuterium exchange in superacids and the reaction of carbenium ions with hydrogen *i.e.* hydrogenolysis by carbenium ions, we suggested that they take place at the H—H bond, *i.e.* they represent alkylation of H—H by alkylcarbenium ions.

In our continued investigations it became obvious that protolysis of single bonds, including C—H as well as C—C bonds is a general reaction[181]. As example, the protolysis of neopentane *(127)* may be mentioned, where there is both C—C and C—H bond reactivity.

In $FSO_3H{-}SbF_5$ or $SbF_5$ (which generally always contains HF) C—H cleavage is favored at low temperatures (−80 °C). At higher temperatures or in $HF{-}SbF_5$, on the other hand, C—C cleavage predominates. This difference could be explained by the differing nature of the solvated proton ($H_2SO_3F^+$ or $H_2F^+$) which in superacids must be considered as a well defined, stable species.

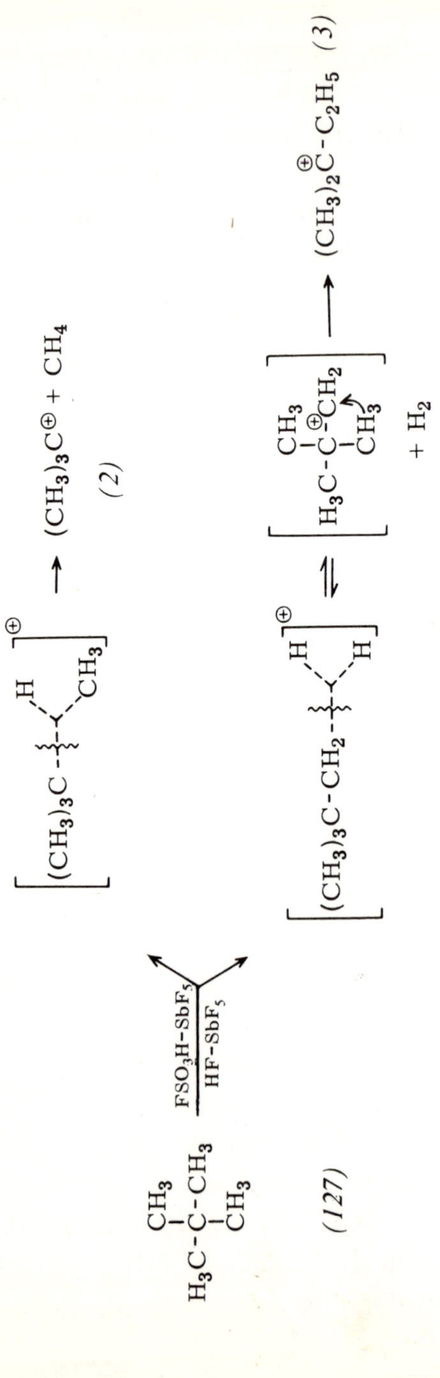

H3C-C(CH3)2-CH3
(127)
FSO3H-SbF5
HF-SbF5
(CH3)3C-[H, CH3]⊕
(CH3)3C⊕ + CH4
(2)
(CH3)3C-CH2-[H, H]⊕
H3C-C(CH3)⊕(CH3)-CH2 + H2
(CH3)2C⊕-C2H5
(3)

As is well known for of n- and π-basicities, steric hindrance can also influence the σ-basicity. If in isobutane the methyl groups are substituted by the much larger *tert*-butyl groups the reactivity of the tertiary C—H bond towards electrophilic reagents decreases rapidly. In the case of tri-*tert*-butylmethane this bond is already shielded to such a high degree that the tertiary C—H bond shows no reactivity at all, and electrophilic reactions, therefore, only occur on the C—C and the primary C—H bonds.

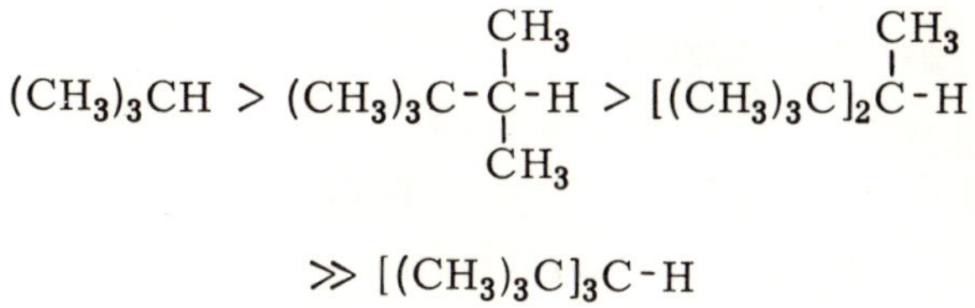

Due to the oxidizing ability of antimony pentafluoride (which is subsequently reduced to $SbF_3$), the possibility of one electron transfer redox processes must always be kept in mind when dealing with systems containing $SbF_5$. In superacids and particularly at low temperature the equilibrium concentration of $SbF_5$, for example in a $HF—SbF_5$ system, is low. Furthermore the protolytic reactivity of alkanes is well demonstrated using deuterated superacids. Electron transfer from an alkane single bond is a relatively high energy process necessitating 4—5 eV (*i. e.* > 100 kcal/mol) more activation energy then a comparable electron transfer process from a related alkene π bond and consequently is not very probable in low temperature solution chemistry. It is worth mentioning that many of the protolytic reactions discussed can also be carried out with acid systems other than those containing

$SbF_5$. HF—$TaF_5$ and HF—$BF_3$ are, for example, useful superacids in such reactions. Since the redox potentials in these systems are high, it seems a well established fact that alkanes react in superacid solutions *via* single bond protolytic processes.

## 4.7. Alkylation

Despite frequent literature references to electrophilic alkylation of alkanes by olefins, from a mechanistic point of view these reactions must be considered as alkylations of the olefin by the carbenium ion formed from the isoalkane by intermolecular hydrogen transfer. Thus the alkane just serves as a reservoir for the alkylating agent.

$$R'_2C{=}CH_2 \xrightarrow{H^\oplus} R'_2\overset{\oplus}{C}{-}CH_3 \xrightarrow{+R_3CH} R'_2CH{-}CH_3 + R_3C^\oplus$$

$$R_3C^\oplus + R'_2C{=}CH_2 \longrightarrow R'_2\overset{\oplus}{C}{-}CH_2{-}CR_3 \xrightarrow{H^\ominus} R'_2CH{-}CH_2{-}CR_3$$

The suggested reaction mechanism (*Schmerling*[189]) is reflected by the products formed in the reaction of propylene and isobutylene with isobutane. The products do not contain 2,2,3-trimethylbutane or 2,2,3,3-tetramethylbutane. These compounds would be expected as primary alkylation products in the direct alkylation of isobutane with propylene and isobutylene, respectively. As already discussed, in intermolecular hydrogen abstraction from a tertiary isoalkane by a carbenium ion, described by *Bartlett*, *Nenitzescu*, and *Schmerling*, the transition state could be considered either linear or triangular (from frontside attack on the C—H bond). The latter would under no circumstances be symmetrical because of steric interaction between the carbenium ion and the tertiary isoalkane. Even with a strongly distorted triangular transition state (*i. e.* assuming that the reaction takes place on the C—H bond itself and not at the back-lobe of the hydrogen atom), it becomes obvious that subsequent cleavage can result not only in intermolecular hydrogen transfer, but also, *via* proton elimination, in direct alkylation.

It should be emphasized that there is no necessity in suggesting a common transition state for hydrogen abstraction and alkylation, but only that triangular states of related nature are formed. As to which products are formed depends on the nature of the reactants, the reaction conditions, and the stability of the products, besides other factors such as strain in the formation of transition states, which affects the ratio of reactions (a) and (b).

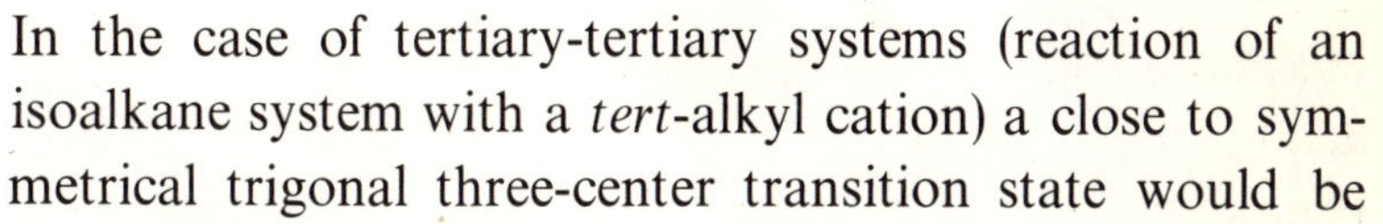

R, R' = Alkyl

In the case of tertiary-tertiary systems (reaction of an isoalkane system with a *tert*-alkyl cation) a close to symmetrical trigonal three-center transition state would be too highly strained for its formation to be possible. However, it is unnecessary to consider a symmetrical transition state in the reactions. The tertiary carbenium ion could easily approach the tertiary C—H bond in such a way as to form an unsymmetric transition state, which can account for the great preference of intermolecular hydrogen transfer over alkylation in the reaction.

The possibility of preparing stable alkylcarbenium ions, permitted these reactions to be investigated under controlled "pure" conditions.

The reaction of stable, alkylcarbenium ions with alkanes can be advantageously studied in sulfuryl chloride fluoride solution at low temperatures (generally as low as −78 °C). A small amount (≈2% of $C_8$ fraction) of 2,2,3,3-tetramethylbutane was detected in the reaction of *tert*-butyl fluoroantimonate with isobutane[180].

$$(CH_3)_3C^{\oplus} + HC(CH_3)_3 \rightleftharpoons (CH_3)_3C{-}C(CH_3)_3 + H^{\oplus}$$

*(128)*

This observation is, however, of substantial importance since no other way than direct alkylation of the C—H bond of isobutane by the *tert*-butyl cation can account for formation of hexamethylethane *(128)*. That direct alkylation of alkanes indeed can be carried out with ease is further shown when isobutane is reacted with the less bulky isopropyl fluoroantimonate, or when propane is reacted with *tert*-butyl fluoroantimonate. The primary alkylation product 2,2,3-trimethylbutane is formed in up to 12% yield[180]. As intermolecular hydrogen transfer is faster than alkylation, and the isopropyl cation more reactive than the *tert*-butyl cation, the alkylation reaction in the system is considered to be mainly the propylation of isobutane.

$$(CH_3)_2CH_2 + \overset{\oplus}{C}(CH_3)_3 \rightleftharpoons (CH_3)_2CH^{\oplus} + HC(CH_3)_3$$
$$(CH_3)_3CH + H\overset{\oplus}{C}(CH_3)_2 \rightleftharpoons (CH_3)_3CCH(CH_3)_2 + H^{\oplus}$$

*(129)*

The alkylation products obtained are not only those derived from the starting alkanes and carbenium ion, but also those from the alkanes and carbenium ions formed by intermolecular hydrogen transfer. Generally this transfer reaction is a faster reaction than alkylation. Since carbocations also undergo isomerization, and since intermolecular hydrogen transfer produces new alkanes from these ions, it is obvious that increasingly more complex mixtures are going to be obtained. Generally the bulkier tertiary carbenium ions can attack the more shielded tertiary C—H bonds only with great difficulty. The processes mentioned, however, readily produce secondary and even incipient primary ions in the systems, which then substitute C—H bonds ("alkylation") or react with C—C bonds ("alkylolysis"). The alkylation reactions of alkanes produce a variety of products which, however, can be quantitatively analyzed by today's methods (gas chromatography, mass spectrometry). Since olefins are formed only to a minimal degree, if at all, under the reaction conditions when stable carbenium ion salts in low nucleophilicity solvents are used and since at low temperatures ($-78\,^{\circ}C$) and short reaction times ($<30$ sec), isomerizations are considered of lesser importance, we feel that indeed, for the first time, we have been able to study direct alkylation of alkanes. Product compositions, which indicate hydrogen transfer as well as primary and secondary alkylation but insignificant olefin formation or isomerization, support this claim. No similar product compositions have been observed in previous acid-catalyzed olefin alkylations. The alkylation of alkanes in competition with simultaneous hydrogen transfer can also be effected with the recently described $CH_3F \rightarrow SbF_5$[191] and $C_2H_5F \rightarrow SbF_5$[192] complexes (which show incipient methyl and ethyl cation properties) and dialkylhalonium ions[193]. It is also reasonable now to suggest that in conventional Friedel-Crafts system direct alkylation can play a role, although in systems where olefin formation is possible, alkene reactivity vastly exceeds that of alkanes.

## 4.8. Nitration

The general concept of electrophilic reactivity of single bonds ($\sigma$ donors) can be further demonstrated with a typical electrophilic reaction such as nitration. To avoid the possibility of any free radical reaction resulting from the use of nitric acid and to avoid acid cleavage of reaction products, nitrations were carried out with stable nitronium salts, such as $NO_2^+PF_6^-$, in aprotic solvents such as methylene chloride-sulfolane[182]. In the case of methane and ethane (where the nitration product—nitromethane and nitroethane—is not acid-sensitive) anhydrous HF or $FSO_3H$ were also used as solvents.

$$R{-}H \xrightarrow[CH_2Cl_2\text{-sulfolane}]{NO_2^{\oplus}PF_6^{\ominus}} R{-}NO_2 + H^{\oplus}$$

So far our studies have been directed primarily to mechanistic aspects. In methylene chloride-sulfolane solutions (which as n-bases compete with hydrocarbon $\sigma$-bases, such as methane) the yields are low (ranging from 1% nitration of methane at 25 °C through 2—5% nitration of higher alkanes, to 10% nitration of adamantane). In HF or $HSO_3F$ solutions much higher yields of nitration of methane and ethane are obtained, but in the case of secondary and tertiary reaction products extensive protolytic cleavage reactions occur in acid media. In the electrophilic aliphatic nitronium ion nitration of alkanes the C—H and C—C bonds are once again the sites of attack; this results in substitution (*i.e.* nitration) or nitrolysis.

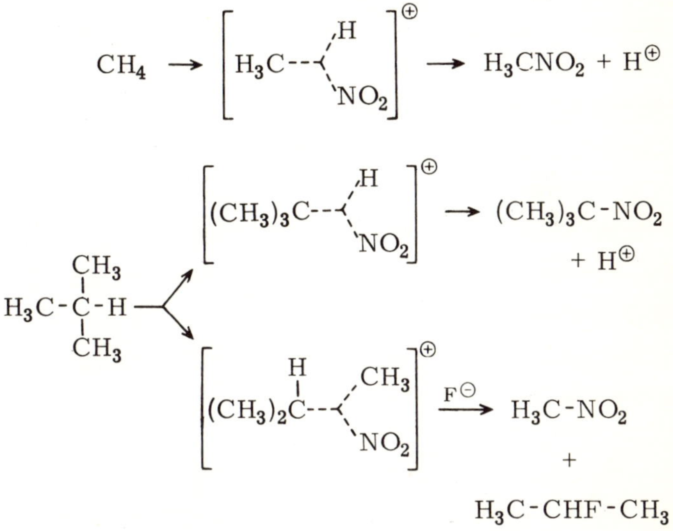

Since tertiary and secondary nitroalkanes are cleaved with ease by strong acids, aliphatic electrophilic nitration is affected by protolytic cleavage of products and therefore generally cannot be carried out in strong acid media without giving complex reaction mixtures.

## 4.9. Chlorination

Chlorination of alkanes can also be carried out under electrophilic conditions, *i.e.* Friedel-Crafts type acid-catalyzed chlorination, which takes place even at −78 °C in the dark[183].

In contrast to other electrophiles, the nature of positive halogens has not yet been sufficiently defined. We were able to carry out the chlorination of methane, ethane, propane, and higher alkanes in $Cl_2$—$SbF_5$—$SO_2ClF$ solution at −78 °C in the dark. In this reaction, substitution products *(chlorination)* are formed as well as C—C cleavage products *(chlorolysis)*. The products formed can once again be explained as resulting from attack of the electrophile on the corresponding bonds with formation of a two-electron three-center bonded transition state. In the reaction of ethane with $Cl_2$—$SbF_5$ in $SO_2ClF$ solution at low temperatures (*i. e.* under conditions favoring stable ions), for example, the main reaction products are the dimethylchloronium ion (chlorolysis) and the diethylchloronium ion (chlorination with subsequent formation of the chloronium ion).

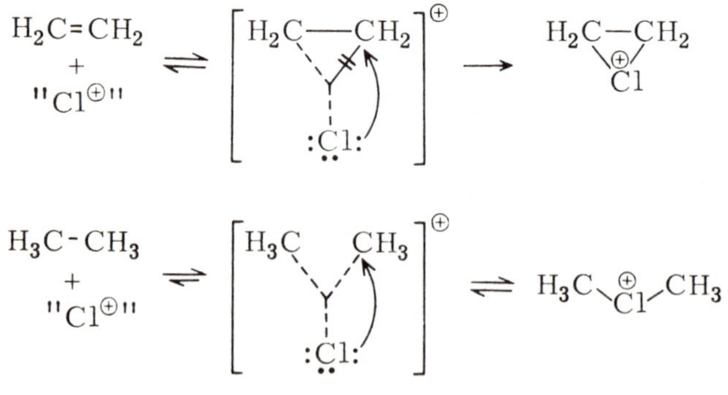

Interestingly, in the case of electrophilic halogenations (such as chlorination) alkanes and alkenes act similarly as electron pair donors. The difference between the paraffins and the olefins in their reactions with "positive" chlorine, in which the unshared electron pairs of chlorine are also involved, is reflected in the type of reaction products formed. In the alkenes, where the σ-bonds are not involved in the reaction, three-membered cyclic halonium (chloronium) ions are formed whereas in the case of alkanes, where the C—C σ-bonds are involved directly in the reaction, open chain dialkyl chloronium ions are formed (chlorolysis).

Alkanes also react in general (*i.e.* under conditions where the intermediate products are unstable) with chlorine in the presence of catalysts such as aluminum chloride, ferric chloride, and antimony pentachloride. Under the mild reaction conditions, these chlorides themselves show no chlorinating ability and are considered to act as true catalysts. (At higher temperatures and in the presence of light or free radical initiators ferric chloride, antimony pentachloride, and phosphorus pentachloride are chlorinating agents.) The reaction conditions used are not considered to favor radical formation, although it must be pointed out that electrophilic chlorine ($Cl^+$ or more probably $Cl_2^+$) can be considered as a radical cation (*i.e.* the triplet state). At the same time, the chlorine atom (Cl·) is a strong electrophile. Hence the differentiation between ionic and radical chlorine is not as clear as in the case of other substituting agents. Further, some catalysts such as $PCl_5$ coordinate weakly with molecular chlorine and could facilitate subsequent homolytic cleavage and thus radical chlorination.

The Friedel-Crafts type chlorination of alkanes affords varying yields (2—5% in the case of methane but up to 55—60% in the chlorination of higher alkanes, including neopentane) and fairly high selectivity. In the chlorination of methane only methyl chloride is obtained (no methylene chloride or chloroform could be detected by gas chromatography). Stronger Lewis acid catalysts ($AlCl_3$, $FeCl_3$, *etc.*) cause increased dehydrochlorination of secondary and tertiary alkyl chlorides with subsequent chlorine addition leading to vicinal dichlorides.

## 4.10. Rearrangements and Related Transformations

With a knowledge of the electron donor character of single bonds which leads to pentacoordinated carbonium ions it is now possible to explain the mechanisms of acid-catalyzed saturated hydrocarbon transformation reactions in those cases where no π- or n-donor reagents are involved which could themselves readily produce the needed initial carbenium ions. In such systems, for example in the isomerization of alkanes, the primary step is protolysis of a C—H or a C—C bond with formation of the corresponding carbonium ion transition state. (We have recently presented a detailed account of the mechanism of protolysis of alkanes[181].) In the following reaction step, with simultaneous cleavage of molecular hydrogen or of an alkane of lower molecular weight, an alkylcarbenium ion is formed which can subsequently undergo intra- or intermolecular processes (alkylation, alkyl or hydrogen shift, hydrogen transfer). These can all be considered to involve interaction of the electron-deficient carbenium center[194] with a C—C or C—H bond with formation of a new carbonium ion transition state. The typical carbocation processes can be shown, for example, in the isomerization of butanes[195].

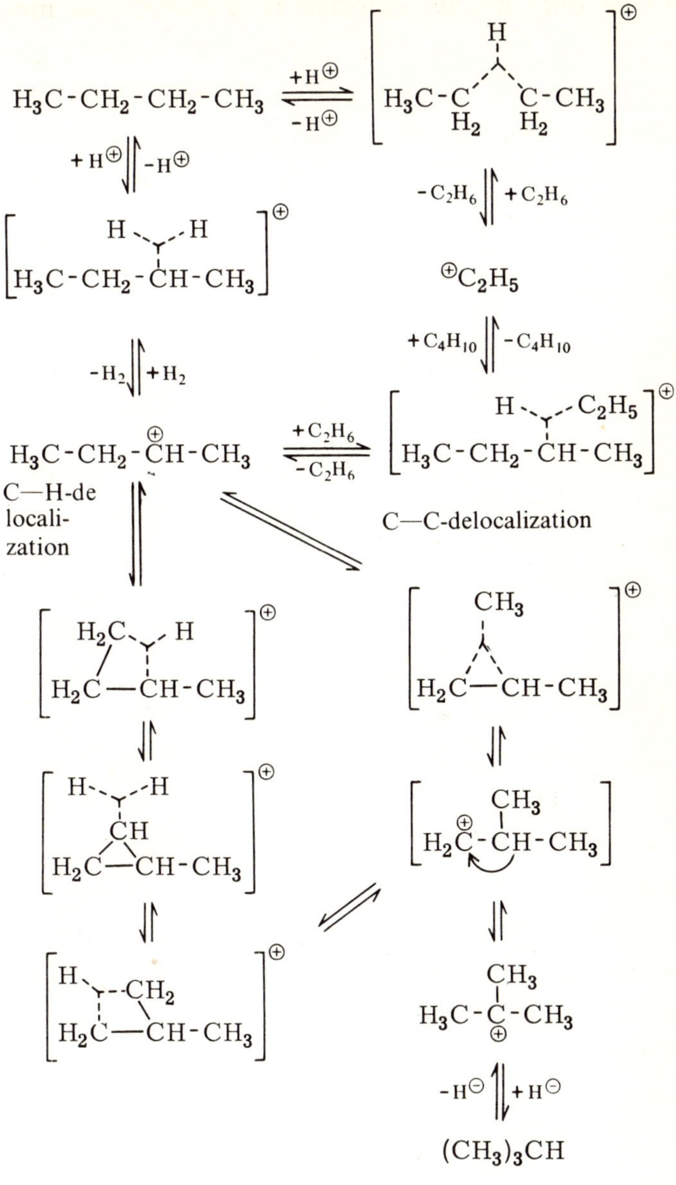

A decisive role is played by protonated cyclopropanes in the rearrangement of the *sec*-butyl cation into the *tert*-butyl cation. The edge- and corner-protonated forms with three-center bonds of type *(a)* and *(b)*, respectively, can be differentiated from the methyl-bridged form [type *(c)*].

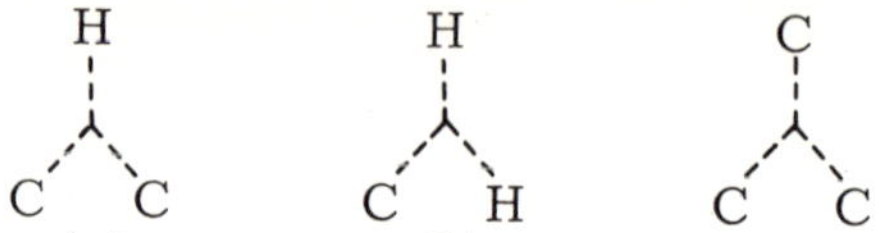


*(a)* *(b)* *(c)*

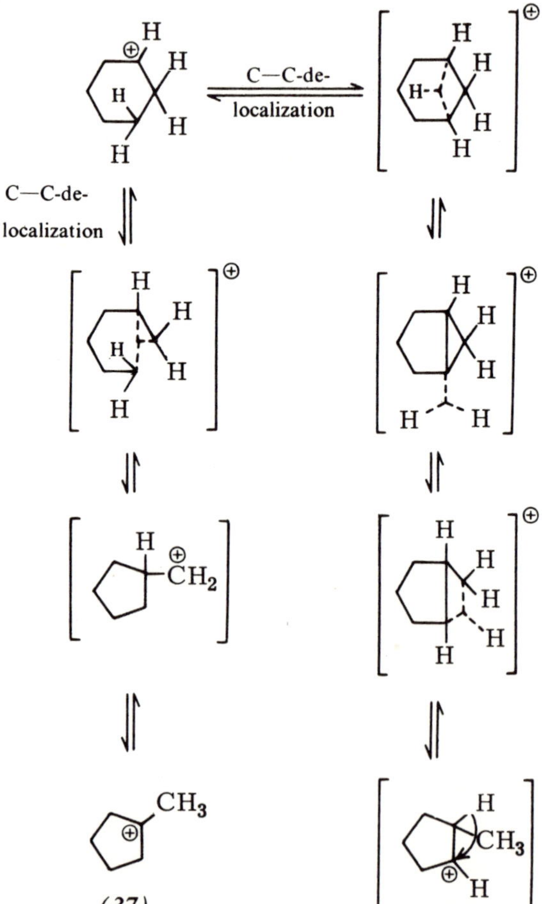


*(37)*

The skeletal rearrangements, however, must always take place with participation of the incipient primary isobutyl cation. Protonated cyclopropanes alone can not account for it, even though they are known to be involved in the hydrogen and carbon exchange (scrambling) processes.

Intramolecular alkylation processes (we have previously discussed well-defined intermolecular alkylations of alkanes by stable alkylcarbenium ions[180]) involving C—C or C—H bonds play a decisive role in ring contraction (or expansion) isomerizations, as, for example, in the isomerization cyclohexane ⇌ methylcyclopentane. Again, either C—H delocalized protonated cyclopropanes or C—C delocalized carbon bridged carbonium ion transition states can be involved.

Whereas the cyclohexane-methylcyclopentane isomerization[196] involves initial formation of the cyclohexyl (methylcyclopentyl) cation (*i.e.* protolysis of a C—H bond) it should be mentioned that in the acid-catalyzed isomerization of cyclohexane up to 10% hexanes are also formed[196, 197]. This is indicative of C—C bond protolysis.

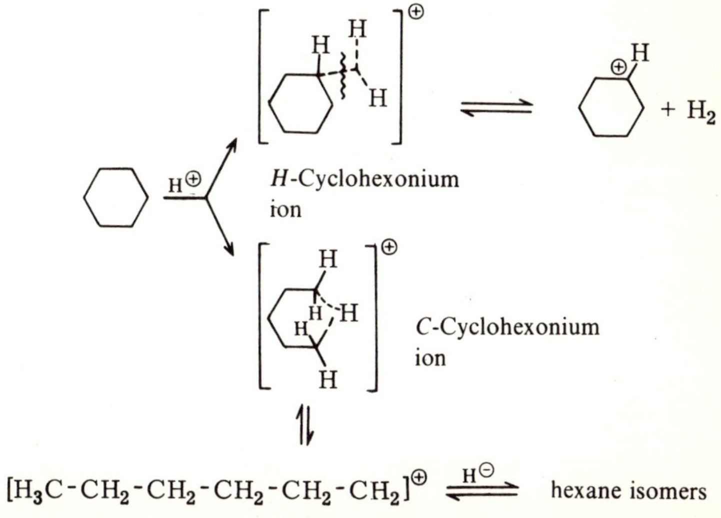

The observed protolytic behavior of cycloalkanes (and polycycloalkanes) in superacid solutions is indicative of either C—H or C—C bond protonation[198].

1,2-$CH_3$(alkyl) or hydrogen shifts (or the similar 1,3-, 1,4- or 1,5-shifts) in alkylcarbenium ions can also be considered as intramolecular alkylation of the C—C or C—H bonds by the carbenium ion center which leads to alkenonium ions. Hydrogen and alkyl bridged alkenonium ions are not observed in superacid solutions of rapidly equilibrating alkylcarbenium ions (*i.e.* by NMR, IR, Raman and ESCA spectroscopy[25, 30, 32, 36]); they are therefore probably only transition states or short-lived intermediates. However, it cannot be decided, as yet, whether alkenonium ion intermediates can exist.

Protolytic or alkylative three-center bond formation in carbonium ion processes can also explain cleavage and elimination reactions, particularly when the cleaving entity is a tertiary carbenium ion or proton; however, secondary and even primary ion cleavages are also observed. The close relationship between substitution, addition, and elimination reactions is thus obvious. Common (or closely related) carbonium ion transition states leading to carbenium ion intermediates are involved, which can be formed either by σ- or π-routes.

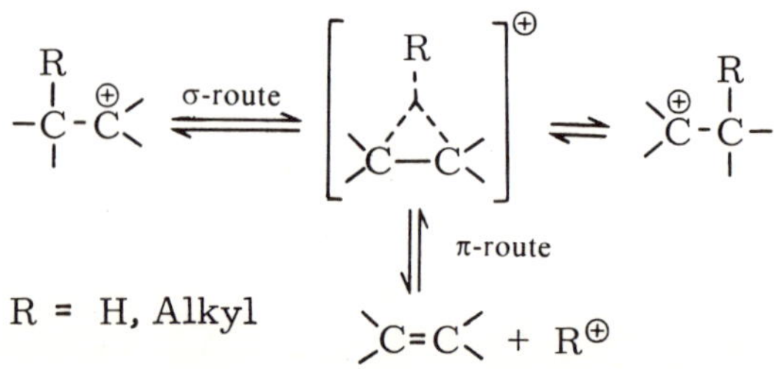

The β-cleavage reactions and deprotolytic olefin formations described by *Whitmore*[199] can also be considered as examples of three-center bound carbonium ion cleavages. *Schleyer*[200], as well as *Brouwer* and *Hogeveen*[201], have called attention to the importance of orientation of adjacent C—H or C—C bonds relative to the vacant p-orbital of carbenium ion centers in intramolecular carbocation reactions (such as 1,2-shifts and β-cleavages). In our opinion the orientation of the vacant p-orbital should be considered with regard to formation of a two-electron three-center bonded carbonium ion transition state.

As already discussed we were recently able to show that the fundamental intermolecular hydride transfer process described by *Bartlett*, *Nenitzescu*, and *Schmerling*[189,190] involves a three-center bound carbonium ion transition state[181] and not, as has been generally assumed, a linear transition state[202]. Even in the sterically highly unfavorable reaction of isobutane with the *tert*-butyl cation, in addition to fast intermolecular hydrogen transfer, formation of small amounts of the alkylation product 2,2,3,3-tetramethylbutane *(129)* was also observed[180]. This

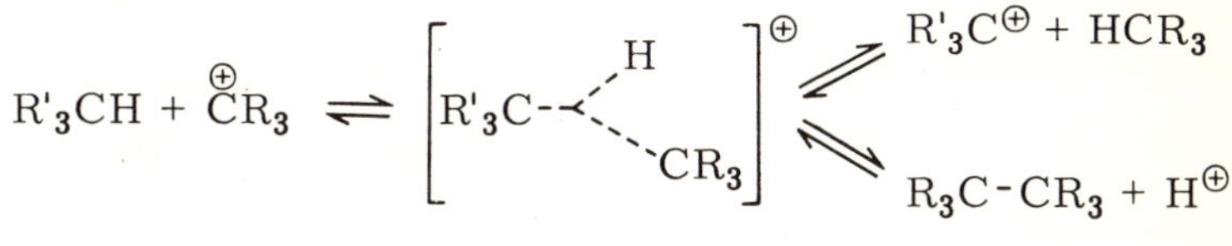


*(129)*, R = $CH_3$

strongly indicates the triangular nature of the transition state. In the frequently used hydrogen abstractions with triphenylmethyl cations[203] (such as in the preparation of tropylium salts from cycloheptatriene[204]) the hydride abstracting agent may not always be the tertiary carbenium ion site. When perdeuteriotriphenylmethyl cation was allowed to react with cycloheptatriene to form tropylium ion, the triphenylmethane formed in the reaction showed protium substitution in the *para* (and *ortho*) ring positions. Hence, not only the tertiary carbenium ion center, but also secondary benzenium ion centers contributing to the stabilization of triarylmethyl cations, participate in the reaction[205].

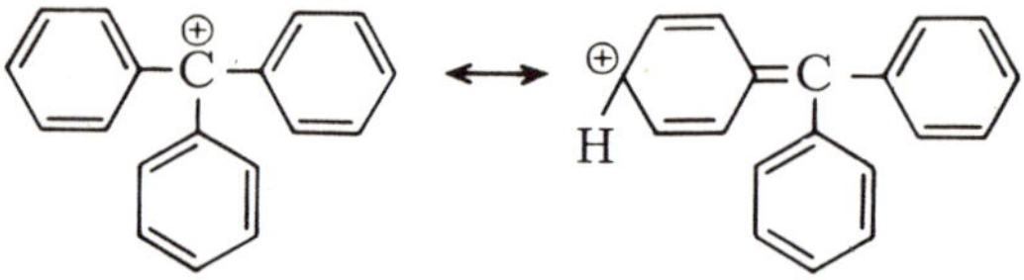


*(40a)* *(40b)*

A somewhat similar situation has emerged from recent reinvestigation of *Gomberg*'s triarylmethyl radicals[206].

## 4.11. The Question of Hyperconjugation

Finally one can raise the question just how much a partially developed intramolecular three-center carbonium ion bond participates in hyperconjugation and the stabilization of carbenium ions. C—H and C—C hyperconjugation are simply explained by this term (where C—C hyperconjugation should play a more important role than C—H hyperconjugation). It should be clear, however, that in three-center bond formation the relative positions of the nuclei involved in this bond (compared to the starting compounds) must change *i.e.* there are indeed new bonding states developed which subsequently can lead to bridging.

In contrast the stabilization of carbenium centers can also take place by σ-π electronic interaction without changes in bond lengths and angles, *i.e.* without extensive motion of the nuclei. This situation was described by *Traylor*[209a] as vertical stabilization (in the Frank-Condon sense). There is, however, no reason to believe that a dichotomy should exist between participation with and without bridging. The hyperconjugative interaction will always be accompanied by some geometrical change, but the extent of that deformation may be small or large (depending on the specific system).

## 4.12. Concerning the Mechanism of Electrophilic Reactions of π-Donor Systems

The reactivity of olefins, acetylenes, and aromatic hydrocarbons toward electrophiles in addition and substitution reactions is based on the π-electron donor ability of the unsaturated C=C or C≡C bonds and the π-aromatic systems[170a, 207].

In the reactions discussed so far we have emphasized the importance of formation of two-electron, three-center bonds in pentacoordinated carbonium ion formation utilizing the electron pair of single σ-bonds. Two-electron, three-center bond formation is also considered to be involved in the initial step of electrophilic reactions of π-bonded systems.

Protonation of olefins, the essential first step in acid-catalyzed isomerizations, alkylation, hydration, and related addition and substitution reactions, is considered to involve the initial overlap of the two carbon atom p-orbitals (of the π-bond) with the empty s-orbital of the proton forming the two-electron, three-center bond of the ethyleneprotonium ion *(130)*, a hydrogen bridged carbonium ion[208].

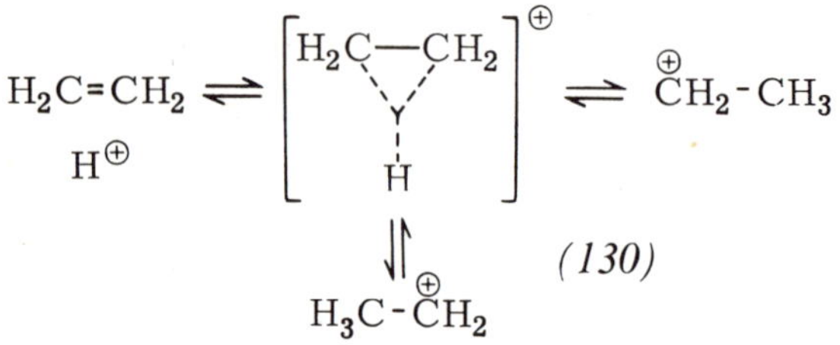

*(130)*

In protonated higher olefins the three-center bond formation should, because of electronic and steric effects, be unsymmetrical (oriented). Furthermore, both theoretical calculations[209] and lack of experimental observation indicate that bridged alkeneprotonium ions (as well as "alkene-alkonium" ions) can exist only as high energy transition states, not as intermediates, but no decision can be made on this point as yet[209]. Subsequent opening of the three-center bond gives the carbenium ion intermediates of highest stability (which explains the empirical Markownikoff rule).

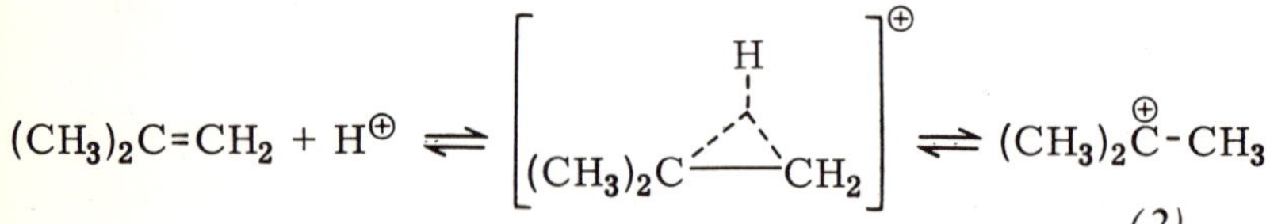

*(2)*

Alkylation, nitration, *etc.* of π-bonds is also considered to involve similar interactions with formation of tetracoordinated carbonium ion transition states *(131)* and *(133)* with three-center bonds.

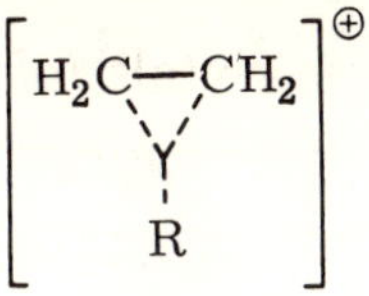

*(131)*

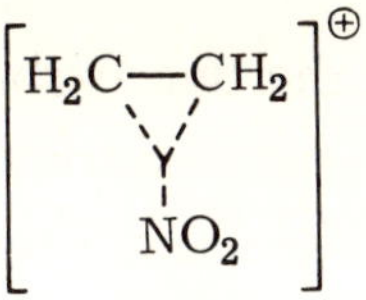

*(132)*

In contrast the long-lived ethylenebromonium ion *(133)* and other alkenehalonium ions are not three-center bound ions, but three-membered cyclic halonium ions with covalent C-halogen bonds[210]. These ions are clearly intermediates and the three-center bound halonium ions such as *(134)*—similar to previously discussed ions—could

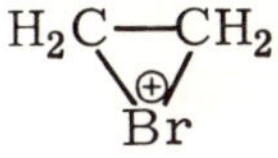

*(133)*

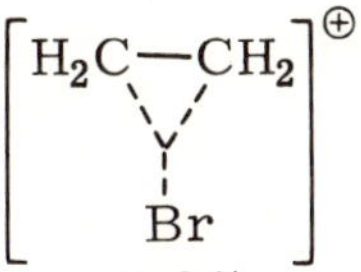

*(134)*

be the transition states leading to them; however, the existence of three-center bonded halonium ion intermediates cannot be ruled out. Distinction between the two different types of alkenonium ions is not always clear,, although it could be, for example, in the case of observable intermediates by comparison of the NMR spectra. NMR data of the observed three-membered ring halonium ions and the three-center bound ethylenemercurinium ion *(135)*[211] (which can serve as model compounds) are quite different. [*(134)*, $\delta_H = 5.53$, $\delta_{^{13}C} = 120.8$; *(135)*, $\delta_H = 7.68$, $\delta_{^{13}C} = 57$.] Even if in the latter case the bonding is somewhat different (due to additional d-orbital participation) to what would be the case in a three-center bound halonium ion, increased p-character of the carbon atomic orbitals in a tetracoordinated carbonium ion should be reflected by substantial deshielding of the chemical shifts (caused also by higher charge density on these atoms).

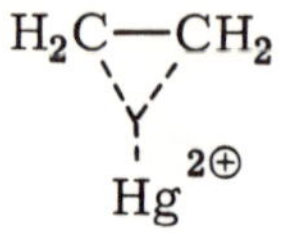

*(135)*

In electrophilic halogen additions to olefins the electrophilic halogen is considered to interact through its vacant atomic orbital with the π-electron pair of the olefin to form a two-electron three-center bonded ion (or probably transition state). Opening of the three-center bond and simultaneous participation of the halogen non-bonded electron pairs then gives the three-membered halonium ion intermediate (which can be spectroscopically observed and also isolated). Subsequent displacement by halide ion from the back-side results in the generally observed *trans* stereospecificity of the addition.

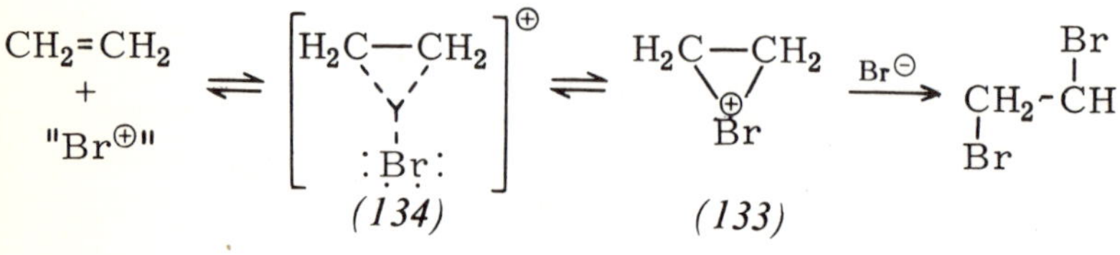

An alternate route can be considered for the initial interaction between olefin and bromine: polarized molecular bromine reacts with the π-bond through overlap of the back-lobe of the antibonding molecular orbital of the bromine to give the three-center bound transition state.

*Strating*, *Wieringa*, and *Wynberg*[212] recently isolated a bromonium ion salt of adamantylideneadamantane.

Recent studies by with *Schilling*, *Westerman*, and *Lin*[212a] by $^1H$ and $^{13}C$-NMR spectroscopy have shown that the $Br_2$ and $Cl_2$ complexes of adamantylideneadamantane, as the $Ag^+$ complex, are indeed three-center bound alkenonium ions (π-complexes), whereas the $H^+$ complex (in superacids) is the rapidly exchanging (equilibrating) alkenium ion (σ-complex). Thus, for the first time a direct differentiation of alkenium (σ-complex) from alkenonium ions (π-complexes) was achieved in electrophilic reactions of alkenes (even if highly hindered ones).

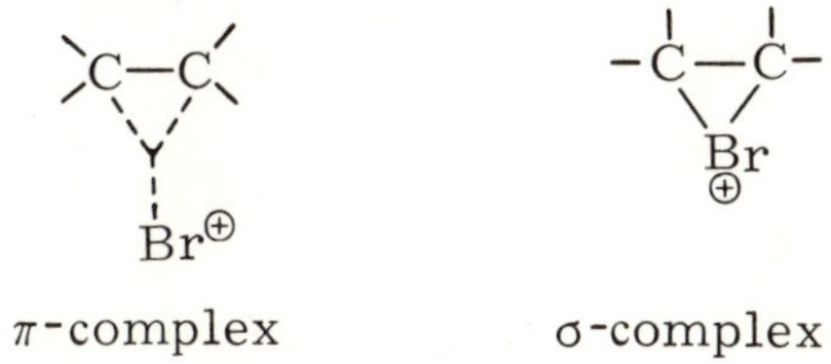


We have also been able to achieve the first direct preparation of cyclic three-membered σ-bound halonium ions by electrophilic halogenation of less hindered olefins by reacting tetramethylethylene and ethylene, respectively, with cyanogen iodide (bromide) in presence of $SbF_5$ in $SO_2$ or $SO_2ClF$ solution at $-78\,^{\circ}C$.

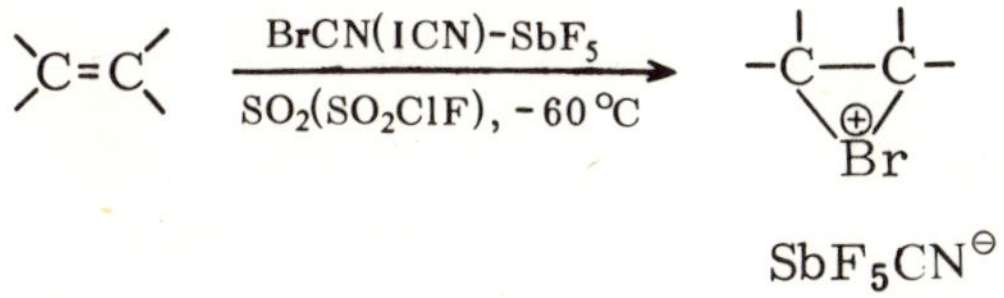


In electrophilic reactions of various π-bonded systems the electrophiles can be defined in a broad sense according to the generalized Lowry-Brønsted concept. Addition reactions, for example, are equally well carried out with various cationic reagents or Lewis acids, such as $BH_3$[213] or $AlH_3$[214] (*Brown* hydroboration[213] or *Ziegler* hydroalanation[214]).

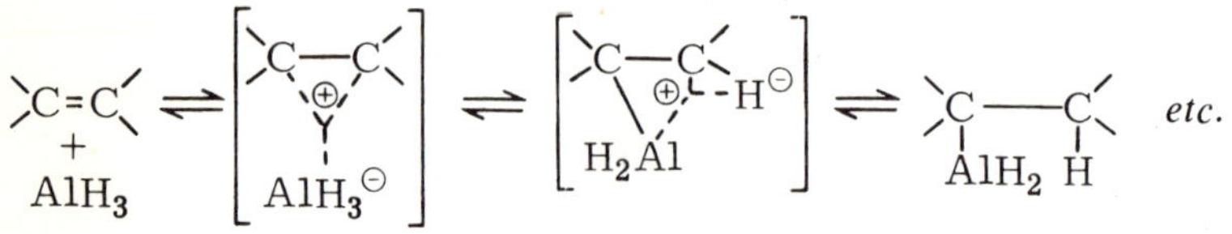

Since the B—H or Al—H bonds act as internal nucleophile during opening of the three-center bond it is clear why hydroboration and hydroalumination show *cis* stereospecificity whereas halogen additions generally proceed by *trans* addition.

Electrophilic reactions by Lewis acid type reagents on σ-bonds are considered less favorable, since the weaker electron donor σ-electron pair is generally not able to form carbonium ions *via* electron pair sharing with Lewis acids (which frequently are dimeric or polymeric halogen bridged reagents like aluminum or antimony halides, and become electron deficient reagents only on breaking the halogen bridges). That is the reason why Friedel-Crafts catalyzed hydrocarbon reactions generally need co-catalysts to provide protic or carbocationic reactivity.

Electrophilic additions to olefins can be considered, in accordance with the Hammond postulate[215], to involve transition states resembling either the intermediate trivalent carbenium (or, in the case of halogenation, three-membered ring halonium) ions or starting olefins. In the first case the new covalent bond is more developed in the transition state than in the latter. Thus the position of the transition state is not fixed and can vary along the reaction path and can resemble either starting olefin or carbenium ion intermediate.

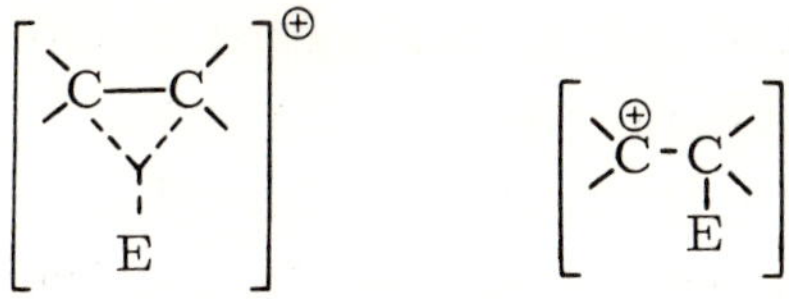

A similar situation is found in aromatic electrophilic substitutions[216]. Once again, in these reactions we must assume the electron donor ability of a π-electron pair towards the electrophile will result in formation of a tetracoordinated three-center bonded benzonium ion *(136)* (π-complex type transition state). Subsequent cleavage of the three-center bond gives a trivalent benzenium ion *(137)* (σ-complex intermediate). The transition state, once again, can either resemble starting aromatics (when it lies early on the reaction coordinate) or be of arenium ion nature (in case of late transition states). For a more detailed discussion of the mechanism of electrophilic aromatic substitution see a recent review [216].

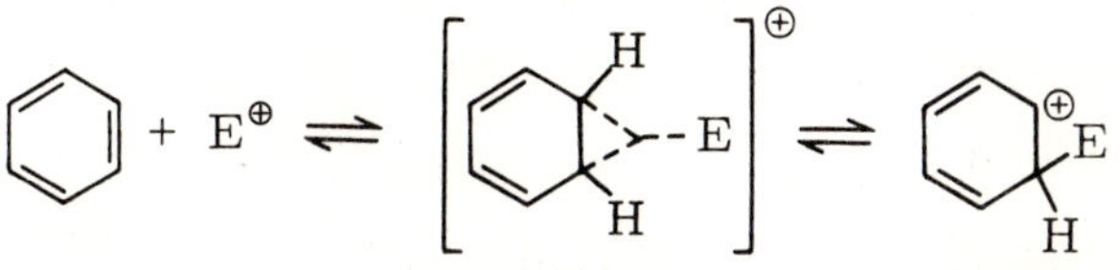

*(136)* *(137)*

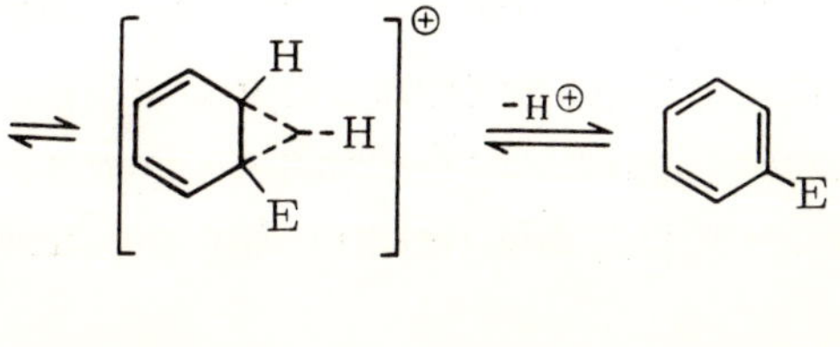

π-Electron pairs are, of course, stronger electron donors than σ-electron pairs. In systems containing both π- and σ-bonds, therefore, the π-system is generally the reactive site. For example, in toluene, electrophilic reactions lead to substitution at the ring and not at the methyl group. However, if we make the π-system less available (either by deactivating groups or by steric crowding) the σ-bonds of the methyl group can become more reactive. Thus pentafluorotoluene undergoes electrophilic reactions on the methyl group[217].

Similar considerations are also valid for the reactivity of deactivated or sterically hindered olefins and other π-donor systems.

*Dewar*[218] should be credited with having advanced the concept of "π-complex" interaction of electrophiles with π-donor systems as far back as in the 1940's. π-Complexes were, however, generally regarded only as weak donor/acceptor interactions between the π-donor system and the electron-deficient reactant. *Dewar*, however, considered them as much stronger "molecular bonds", the π-electron pair overlapping with the electron-deficient species' empty orbital. In 1965, when discussing "nonclassical ions" as π-complexes, *Dewar* and *Marchand*[219] stated, "...in a true π-complex, the π-donor and acceptor are linked by a covalent bond, not by weak forces of van der Waal's type." Dewar's π-complexes indeed are carbonium ions (alkonium ions) involving two-electron, three-center bonds, with the π-system supplying the electron pair for the bond.

It should be emphasized that the chemical literature contains in many instances suggestions for bridged (symmetrical or unsymmetrical) cationic intermediates, besides

Dewar's π-complexes. For example, *Taft*[208] suggested that the protonation of olefins leads first to a separate π-bonded species which subsequently can transform into a carbenium ion. *Ingold*[220], *Winstein*[221], *Bartlett*[222], *Cram*[223], *Stiles*[224], *Berson*[225], *Allen*, *Schleyer*, and *Pople*[226] as well as other investigators have on different occasions discussed in considerable depth the question of bridging in carbocations. Only our recent studies, however, have shown that the unifying principle of all electrophilic reactions is indeed the general donor ability of the bonded electron-pairs of either π- or σ-bonds to form, with the vacant orbital of the electrophile, triangular two-electron three center bound carbonium ions (transition states or intermediates), which subsequently can give trivalent carbenium ions or lead to substitution (addition) products[12].

## 5. Outlook

In conclusion, the differentiation of trivalent carbenium from penta- or tetracoordinated carbonium ions, based on their direct observation, showing marked differences, for example, in their NMR and ESCA spectra, should put an end to much of the "myth" surrounding organic cations and to the so-called "classical-nonclassical ion controversy". In defining limiting categories of carbocations it must be clearly understood, of course, that there exists the possibility of a whole spectrum of ions or intermediate degrees of delocalization (or partial "nonclassical" nature) and therefore division in strictly limiting categories frequently is arbitrary. Whereas the norbornyl cation itself, for example, is a completely symmetrically delocalized nonclassical carbonium ion, the 2-methylnorbornyl cation shows partial delocalization. More importantly, the concept of pentacoordinated carbonium ion formation *via* electron sharing of single bonds with electrophilic reagents in three-center bond formation promises to open up a whole new important area of chemistry. Whereas the concept of tetravalency of carbon obviously is not affected, carbon penta- (or tetra-)coordination as a general phenomenon must be recognized. Trivalent carbenium ions play a major role in electrophilic reactions of π- and n-donors, whereas pentacoordinated carbonium ions play an equally important similar role in electrophilic reactions of σ-donor saturated systems. Accordingly electron donors can be categorized into the following types:

Types of electron donors

| π-Donors | σ-Donors | n-Donors |
|---|---|---|
| bonded (shared) electron pair donors | | non-bonded (unshared) electron pair donors |
| alkenes<br>alkynes<br>aromatics | alkanes<br>single bonds of all types, such as C—H, C—C, H—H, *etc.* | hetero-atoms, such as O, S, N, Hal *etc.*<br>carbanions |

The realization of the electron donor ability of shared (bonded) electron pairs (single bonds) could one day rank equal in importance with *G. N. Lewis*' realization[227] of the importance of the electron donor unshared (non-bonded) electron pairs. We can now not only explain the reactivity of saturated hydrocarbons and single bonds in general in electrophilic reactions, but indeed use this understanding to explore many new areas and reactions of carbocation chemistry.

Finally, I would like to point out that my review deals primarily with the structural problems of carbocations based on the direct (spectroscopic) observation of long-lived ions and the role carbocations play in the course of electrophilic reactions. No attempt has been made to give a complete survey of this very large and active field. I would, therefore, like to apologize to all those colleagues working in this field whose important contributions have not been mentioned. The choice of material was strongly influenced by limitations of space and my own interests and should not be assumed as a critical evaluation. Our own studies were inevitably strongly emphasized and served as a basis for development of the general concepts presented.

*Support of our work by the National Science Foundation, the Petroleum Research Fund administered by the American Chemical Society, the U.S. Army Research Office, and the National Institutes of Health, US Public Health Service is gratefully acknowledged. My particular thanks go to all members of my research group (past and present) whose enthusiastic hard work really made our studies possible. Their names are found in the references.*

*The John Simon Guggenheim Memorial Foundation is thanked for a fellowship which greatly facilitated the preparation of this manuscript.*

## References

[1] Stable Carbocations Part 138.—Part 137: *G. A. Olah* and *G. Liang*, J. Amer. Chem. Soc. *94*, 6434 (1972).

[2] For reviews see a) *G. A. Olah* and *P. v. R. Schleyer:* Carbonium Ions. Wiley-Interscience, New York, Vol. 1, 1968, Vol. 2, 1970, Vol. 3, 1972, Vol. 4, 1973; b) *D. Bethell* and *V. Gold:* Carbonium Ions, An Introduction. Academic Press, New York 1967.

[3] For reviews see *G. A. Olah*, Chem. Eng. News *45*, p. 76 (March 27, 1967); Science *168*, 1298 (1970).

[4] Existence is defined by Webster (Third New International Dictionary, Unabridged Edit., Merriam Co., 1961, p. 796) as: "The state or fact of having being, especially as considered independently of human consciousness."

[5] *R. J. Gillespie*, J. Chem. Soc. *1952*, 1002; and for a rebuttal, *M. J. S. Dewar*, *ibid. 1953*, 2885; *H. H. Jaffe*, J. Chem. Phys. *21*, 1893 (1953); but also *R. J. Gillespie*, *ibid. 21*, 1893 (1953).

[6] *E. L. Muetterties* and *R. A. Schunn*, Quart. Rev. Chem. Soc. *20*, 245 (1966).

[7] For a review and critical discussion, see *F. R. Jensen* and *B. Rickborn:* Electrophilic Substitution of Organomercurials. McGraw-Hill, New York 1968.

[8] *Roberts* and *Winstein* seem to have been the first to use the term "nonclassical ion"—*Roberts* when he proposed the tricyclobutonium structure for the cyclopropylmethyl cation [*J. D. Roberts* and *R. H. Mazur*, J. Amer. Chem. Soc. *73*, 1542 (1951)], and *Winstein* when he referred to the nonclassical structure of norbornyl, cholesteryl, and 3-phenyl-2-butyl cations [*S. Winstein* and *D. Trifan*, *ibid. 74*, 1154 (1952)]. *Bartlett*'s definition is widely used for nonclassical ions: "an ion is nonclassical if its ground state has delocalized bonding σ-electrons" [9a].

[9] For leading references see: a) *P. D. Bartlett:* Nonclassical Ions. Reprints and Commentary. Benjamin, New York 1965; b) *G. D. Sargent*, Quart. Rev. Chem. Soc. *20*, 301 (1966); c) *S. Winstein*, *ibid. 23*, 141 (1969); and for an opposing view d) *H. C. Brown*, Chem. Eng. News, *45*, p. 86 (Feb. 13, 1967), and references given therein.

[10] *G. A. Olah*, *A. M. White*, *J. R. DeMember*, *A. Commeyras*, and *C. Y. Lui*, J. Amer. Chem. Soc. *92*, 4627 (1970), and references given therein.

[11] For a review and discussion, see *J. J. Jennen*, Chimia *20*, 309 (1966).

[12] *G. A. Olah*, J. Amer. Chem. Soc. *94*, 808 (1972).

[13] *H. Meerwein* and *K. Van Emster*, Chem. Ber. *55*, 2500 (1922).

[14] For a summary see *C. K. Ingold:* Structure and Mechanism in Organic Chemistry. Cornell Univ. Press, Ithaca, N. Y., 1953.

[15] *R. C. Whitmore*, J. Amer. Chem. Soc. *54*, 3274, 3276 (1932); Ann. Rep. Progr. Chem. (Chem. Soc. London) *1933*, 177.

[16] For an extensive review see *F. W. McLafferty:* Mass Spectrometry of Organic Ions. Academic Press, New York 1963.

[17] *H. C. Brown*, *H. W. Pearsall*, and *L. P. Eddy*, J. Amer. Chem. Soc. *72*, 5347 (1950).

[18] *E. Wertyporoch* and *T. Firla*, Ann. Chim. *500*, 287 (1933).

[19] *G. A. Olah*, *S. J. Kuhn*, and *J. A. Olah*, J. Chem. Soc. *1957*, 2174.

[20] *F. Fairbrother*, J. Chem. Soc. *1945*, 503.

[21] *J. Rosenbaum* and *M. C. R. Symons*, Proc. Chem. Soc. (London) *1959*, 92; *J. Rosenbaum* and *M. C. R. Symons*, Mol. Phys. *3*, 205 (1960); *J. Rosenbaum* and *M. C. R. Symons*, J. Chem. Soc. *1961*, 1.

[22] *A. C. M. Finch* and *M. C. R. Symons*, J. Chem. Soc. *1965*, 378.

[23] For a summary see *N. C. Deno*, Progr. Phys. Org. Chem. *2*, 129 (1964).

[24] For preliminary communications and lectures, see: *G. A. Olah*, Conference Lecture at the 9th Reaction Mechanism Conference, Brookhaven, New York, August 1962; Abstr. 142nd Nat. Meet. Amer. Chem. Soc., Atlantic City, N. J. 1962, p. 45; *G. A. Olah*, *W. S. Tolgyesi*, *J. S. MacIntyre*, *I. J. Bastien*, *M. W. Meyer*, and *E. B. Barker*, Abstr. A, 19th Int. Congr. Pure Appl. Chem., London, 1963, p. 121; *G. A. Olah*, Angew. Chem. *75*, 800 (1963); Rev. Chim. Acad. Rep. Populaire Roumaine *7*, 1139 (1962); Prepr. Div. Petrol. Chem. Amer. Chem. Soc. *9* (No. 2), C 31 (1964); Organic Reaction Mechanism Conference, Cork, Ireland, June 1964, Spec. Publ. No. 19, Chemical Society, London 1965.

[25] *G. A. Olah*, *E. B. Baker*, *J. C. Evans*, *W. S. Tolgyesi*, *J. S. McIntyre*, and *I. J. Bastien*, J. Amer. Chem. Soc. *86*, 1360 (1964).

[26] *G. A. Olah*, *W. S. Tolgyesi*, *S. J. Kuhn*, *M. E. Moffatt*, *I. J. Bastien*, and *E. B. Baker*, J. Amer. Chem. Soc. *85*, 1328 (1963).

[27] a) *J. Bacon*, *P. A. W. Dean*, and *R. J. Gillespie*, Can. J. Chem. *47*, 1655 (1969); b) *A. Commeyras* and *G. A. Olah*, J. Amer. Chem. Soc. *91*, 2929 (1969).

[28] Superacids are, according to *R. J. Gillespie* [Accounts Chem. Res. *1*, 202 (1968)], more acidic than 100% sulfuric acid, the most frequently

used strong acid solvent. The name superacid was first suggested by *Conant* in 1927 [*N. F. Hall* and *J. B. Conant*, J. Amer. Chem. Soc. *49*, 3047 (1927)] for nonaqueous strong acid solutions (like perchloric acid solution) of weak bases (like amides) where salt formation is exceptionally complete.

[29] Magic acid® [27b] is a registered trade mark of Cationics, Inc. 653 Alpha Drive, Cleveland, Ohio 44143 (USA) and is available, together with other superacids and cationic reagents, from this company as well as from Aldrich Chemical Co.

[30] For a comprehensive report see *G. A. Olah* and *A. M. White*, J. Amer. Chem. Soc. *91*, 5801 (1969).

[30a] For an account of INDOR spectroscopy see *W. von Philipsborn*, Angew. Chem. *83*, 470 (1971); Angew. Chem. internat. Edit. *10*, 472 (1971).

[31] *M. Saunders* and *E. L. Hagen*, J. Amer. Chem. Soc. *90*, 2436 (1968).

[31a] *G. J. Karabatsos* and *C. E. Orzech*, J. Amer. Chem. Soc. *86*, 3574 (1964).

[32] *G. A. Olah, A. Commeyras, J. DeMember*, and *J. L. Bribes*, J. Amer. Chem. Soc. *93*, 459 (1971).

[33] *G. A. Olah, C. U. Pittman, Jr., R. Waack*, and *M. Doran*, J. Amer. Chem. Soc. *88*, 1488 (1966).

[34] *G. A. Olah, C. U. Pittman, Jr.*, and *M. C. R. Symons*, in *G. A. Olah* and *P. v. R. Schleyer*, Carbonium Ions. Wiley-Interscience, New York 1968, Vol. 1, pp. 153ff.

[35] a) *K. Siegbahn, C. Nordling, A. Fahlman, R. Nordberg, K. Hamrin, J. Hedman, G. Johansson, T. Bergmark, S. E. Karlsson, I. Lindgren*, and *B. Lindberg:* ESCA, Atomic, Molecular and Solid State Structure Studies by Means of Electron Spectroscopy. Almquist & Wiksells Boktryckeri AB, Uppsala 1967; b) for reviews see *J. J. M. Hollander* and *W. L. Jolly*, Accounts Chem. Res. *3*, 193 (1970); *D. Betteridge* and *A. D. Baker*, Anal. Chem. *42*, 43A (1970); *D. M. Hercules, ibid. 42*, 20A (1970); c) cf. also *C. Nordling*, Angew. Chem. *84*, 144 (1972); Angew. Chem. internat. Edit. *11*, 83 (1972).

[36] a) *G. A. Olah, G. D. Mateescu, L. A. Wilson*, and *M. H. Goss*, J. Amer. Chem. Soc. *92*, 7231 (1970); b) *G. A. Olah, G. D. Mateescu*, and *J. L. Riemenschneider, ibid. 94*, 2529 (1972).

[37] *G. A. Olah*, Chem. Eng. News *45*, p. 76 (March 27, 1967); *G. A. Olah* and *Y. Halpern*, J. Org. Chem. *36*, 2354 (1971).

[38] *D. M. Brouwer* and *E. L. Mackor*, Proc. Chem. Soc. (London) *1964*, 147; *D. M. Brouwer*, Rec. Trav. Chim. Pays-Bas *87*, 210 (1968).

[39] *G. A. Olah, M. B. Comisarow, C. A. Cupas,* and *C. U. Pittman, Jr.*, J. Amer. Chem. Soc. *87*, 2997 (1965).

[40] *G. A. Olah* and *E. Namanworth*, J. Amer. Chem. Soc. *88*, 5327 (1966).

[41] *G. A. Olah, J. Sommer;* and *E. Namanworth*, J. Amer. Chem. Soc. *89*, 3576 (1967).

[42] *G. A. Olah, J. M. Bollinger,* and *P. Schilling*, to be published.

[43] *G. A. Olah* and *D. H. O'Brien*, J. Amer. Chem. Soc. *89*, 1725 (1967).

[44] *G. A. Olah, D. H. O'Brien,* and *C. U. Pittman, Jr.*, J. Amer. Chem. Soc. *89*, 2996 (1967).

[45] *G. A. Olah* and *J. Lukas*, J. Amer. Chem. Soc. *89*, 2227, 4739 (1967).

[46] *H. Hogereen* and *C. J. Gaasbeck*, Rec. Trav. Chim. Pays-Bas *89*, 395 (1970), and literature cited therein.

[47] *G. A. Olah, D. H. O'Brien,* and *C. U. Pittman, Jr.*, J. Amer. Chem. Soc. *89*, 2996 (1967).

[48] *G. A. Olah* and *P. Schilling*, Liebigs Ann. Chem. *761*, 77 (1972).

[49] *G. A. Olah, N. Friedman, J. M. Bollinger,* and *J. Lukas*, J. Amer. Chem. Soc. *88*, 5328 (1966).

[50] *J. D. Roberts* and *R. H. Mazur*, J. Amer. Chem. Soc. *73*, 2509 (1951).

[51] *H. Hart* and *J. M. Sandri*, J. Amer. Chem. Soc. *81*, 320 (1959).

[52] *N. C. Deno, H. G. Richey, Jr., S. Liu, J. D. Hodge, J. J. Houser,* and *M. J. Wisotsky*, J. Amer. Chem. Soc. *84*, 2016 (1962).

[53] *G. A. Olah* and *P. Westerman*, to be published.

[54] *C. U. Pittman, Jr.*, and *G. A. Olah*, J. Amer. Chem. Soc. *87*, 2998 (1965).

[54a] *D. S. Kabakoff* and *E. Namanworth*, J. Amer. Chem. Soc. *92*, 3234 (1970).

[55] *N. C. Deno, H. G. Richey, Jr., N. Friedman, J. D. Hodge, J. J. Houser,* and *C. U. Pittman, Jr.*, J. Amer. Chem. Soc. *85*, 2991 (1963); *N. C. Deno, J. M. Bollinger, N. Friedman, K. Hafer, J. D. Hodge,* and *J. J. Houser, ibid.* *85*, 2998 (1963).

[56] *T. S. Sorensen*, Can. J. Chem. *42*, 2768 (1969).

[57] *G. A. Olah* and *M. B. Comisarow*, J. Amer. Chem. Soc. *86*, 5682 (1964).

[58] *G. A. Olah* and *J. M. Bollinger*, J. Amer. Chem. Soc. *90*, 6082 (1968).

[59] *G. Biale, A. J. Parker, S. G. Smith, I. D. R. Stevens*, and *S. Winstein*, J. Amer. Chem. Soc. *92*, 115 (1970).

[60] *P. v. R. Schleyer, T. M. Su, M. Saunders*, and *J. C. Rosenfeld*, J. Amer. Chem. Soc. *91*, 5174 (1969).

[61] *J. M. Bollinger, J. M. Brinich*, and *G. A. Olah*, J. Amer. Chem. Soc. *92*, 4025 (1970).

[62] *G. A. Olah* and *J. M. Bollinger*, J. Amer. Chem. Soc. *90*, 6083 (1968).

[63] *C. U. Pittman, Jr.*, Chem. Commun. *1969*, 122.

[64] *T. S. Sorensen*, J. Amer. Chem. Soc. *87*, 5075 (1965); Can. J. Chem. *43*, 2744 (1965).

[65] *G. A. Olah, C. U. Pittman, Jr.*, and *T. S. Sorensen*, J. Amer. Chem. Soc. *88*, 2331 (1966).

[66] *A. A. Verrijn Stuart* and *E. L. Mackor*, J. Chem. Phys. *27*, 826 (1957); *G. Dallinga, E. L. Mackor*, and *A. A. Verrijn Stuart*, Mol. Phys. *1*, 123 (1968); *C. Maclean* and *E. L. Mackor*, Discuss. Faraday Soc. *34*, 165 (1962).

[67] *T. Birchall* and *R. J. Gillespie*, Can. J. Chem. *42*, 502 (1964).

[68] *G. A. Olah* and *S. J. Kuhn*, Nature *178*, 693 (1956); Naturwissenschaften *43*, 59 (1956); J. Amer. Chem. Soc. *80*, 6535 (1958); *G. A. Olah*, Abstr. 138th Nat. Meet. Amer. Chem. Soc., New York 1960, p. 3P; J. Amer. Chem. Soc. *87*, 1103 (1965).

[69] *G. A. Olah, R. H. Schlosberg, R. D. Porter, Y. K. Mo, D. P. Kelly*, and *G. D. Mateescu*, J. Amer. Chem. Soc. *94*, 2034 (1972).

[70] *W. v. E. Doering, M. Saunders, H. G. Boyton, H. W. Earhart, E. F. Wadley, W. R. Edwards*, and *G. Laber*, Tetrahedron *4*, 178 (1958).

[71] *G. A. Olah, G. D. Mateescu*, and *Y. K. Mo*, J. Amer. Chem. Soc., in press.

[72] *G. D. Mateescu*, Ph. D. Dissertation, Case Western Reserve University 1971.

[73] *G. A. Olah, J. M. Bollinger, C. A. Cupas*, and *J. Lukas*, J. Amer. Chem. Soc. *89*, 2692 (1967).

[74] *D. M. Brouwer, E. L. Mackor*, and *C. MacLean* in *G. A. Olah* and *P. v. R. Schleyer:* Carbonium Ions. Wiley-Interscience, New York 1970, Vol. 2, Chap. 20, and literature cited therein.

[75] *G. A. Olah* and *J. Lukas*, J. Amer. Chem. Soc. *90*, 933 (1968).

[76] *G. A. Olah, C. L. Jeuell, D. P. Kelly*, and *R. D. Porter*, J. Amer. Chem. Soc. *94*, 146 (1972).

[77] *J. Bredt, J. Houben,* and *P. Levy,* Chem. Ber. *35,* 1286 (1902); *J. Bredt, H. Thouet,* and *J. Schmitz,* Liebigs Ann. Chem. *437,* 1 (1924).

[78] *P. D. Bartlett* and *L. H. Knox,* J. Amer. Chem. Soc. *61,* 3184 (1939).

[79] *H. Koch* and *W. Haaf,* Angew. Chem. *72,* 628 (1960).

[80] *P. v. R. Schleyer, W. E. Watts, R. C. Fort, Jr., M. B. Comisarow,* and *G. A. Olah,* J. Amer. Chem. Soc. *86,* 4195 (1964).

[81] *G. A. Olah, G. D. Mateescu, G. Sipos,* and *P. v. R. Schleyer,* J. Amer. Chem. Soc., in press.

[82] *P. v. R. Schleyer, W. E. Watts, M. B. Comisarow,* and *G. A. Olah,* J. Amer. Chem. Soc. *86,* 5679 (1964).

[83] *J. F. Norris,* Amer. Chem. J. *25,* 117 (1901); *J. F. Norris* and *W. W. Sanders, ibid. 25,* 54 (1901).

[84] *F. Kehrmann* and *F. Wentzel,* Chem. Ber. *34,* 3815 (1901); *35,* 622 (1902).

[85] *A. Baeyer* and *V. Villiger,* Chem. Ber. *35,* 1189, 3013 (1902).

[86] *V. Gold* and *F. L. Tye,* J. Chem. Soc. *1952,* 2172.

[87] *D. G. Farnum,* J. Amer. Chem. Soc. *86,* 934 (1964).

[88] *N. C. Deno, E. Bocker, C. U. Pittman, Jr.,* and *J. O. Turner,* Progr. Phys. Org. Chem. *2,* 129 (1964), and literature cited therein.

[89] *G. A. Olah,* J. Amer. Chem. Soc. *86,* 932 (1964).

[90] *G. A. Olah* and *C. U. Pittman, Jr.,* J. Amer. Chem. Soc. *87,* 3507 (1965); *G. A. Olah, C. U. Pittman, Jr., E. Namanworth,* and *M. B. Comisarow, ibid. 88,* 5571 (1966).

[91] *C. A. Cupas, M. B. Comisarow,* and *G. A. Olah,* J. Amer. Chem. Soc. *88,* 361 (1966).

[92] *J. M. Bollinger, M. B. Comisarow, C. A. Cupas,* and *G. A. Olah,* J. Amer. Chem. Soc. *89,* 5687 (1967).

[93] *G. A. Olah, R. D. Porter,* and *D. P. Kelly,* J. Amer. Chem. Soc. *93,* 464 (1971).

[94] *H. Hart* and *R. W. Fish,* J. Amer. Chem. Soc. *80,* 5894 (1958); *82,* 5419 (1960); *83,* 4460 (1961).

[95] *R. J. Gillespie* and *E. A. Robinson,* J. Amer. Chem. Soc. *86,* 5676 (1964).

[96] *N. C. Deno, N. Friedman,* and *J. Mockus,* J. Amer. Chem. Soc. *86,* 5676 (1964).

[97] *E. A. Robinson* and *J. A. Ciruna,* J. Amer. Chem. Soc. *86,* 5677 (1964).

[98] *H. H. Freedman* and *A. M. Frantz, Jr.*, J. Amer. Chem. Soc. *84*, 4165 (1962).

[99] *R. F. Bryan*, J. Amer. Chem. Soc. *86*, 733 (1964).

[100] *G. A. Olah, J. M. Bollinger*, and *A. M. White*, J. Amer. Chem. Soc. *91*, 3667 (1969).

[101] *G. A. Olah* and *G. D. Mateescu*, J. Amer. Chem. Soc. *92*, 1430 (1970).

[102] *R. Breslow* and *C. Yuan*, J. Amer. Chem. Soc. *80*, 5991 (1958); *R. Breslow, J. T. Groves*, and *G. Ryan, ibid. 89*, 5048 (1967).

[103] *H. Hart, T. Sulzberg*, and *R. R. Rafos*, J. Amer. Chem. Soc. *85*, 1800 (1963).

[104] *H. Volz* and *M. J. Volz de Lecea*, Tetrahedron Lett. *1964*, 1871; *1966*, 4863.

[105] *J. M. Bollinger, C. A. Cupas, K. J. Friday, M. L. Woolfe*, and *G. A. Olah*, J. Amer. Chem. Soc. *89*, 156 (1967).

[106] *G. Merling*, Chem. Ber. *24*, 3108 (1891).

[107] *W. v. E. Doering* and *L. H. Knox*, J. Amer. Chem. Soc. *76*, 3203 (1954).

[108] *H. J. Dauben, Jr., F. A. Gadecki, K. M. Harmon*, and *D. L. Person*, J. Amer. Chem. Soc. *79*, 4557 (1957).

[109] *G. A. Olah, M. B. Comisarow*, and *C. A. Cupas*, J. Amer. Chem. Soc. *88*, 362 (1966).

[110] For a summary see a) *G. A. Olah* and *Y. K. Mo*, Advan. Fluorine Chem., in press; b) *G. A. Olah* and *Y. K. Mo* in *G. A. Olah* and *P. v. R. Schleyer:* Carbonium Ions. Wiley-Interscience, New York, Vol. 5, in press.

[111] *G. A. Olah, R. D. Chambers*, and *M. B. Comisarow*, J. Amer. Chem. Soc. *89*, 1268 (1967).

[112] *G. A. Olah* and *C. U. Pittman, Jr.*, J. Amer. Chem. Soc. *88*, 3310 (1966).

[113] *G. A. Olah* and *M. B. Comisarow*, J. Amer. Chem. Soc. *89*, 1027 (1967).

[114] *R. Filler, C. S. Wang, M. A. McKinney*, and *F. N. Miller*, J. Amer. Chem. Soc. *89*, 1026 (1967).

[115] *P. B. Sargent* and *C. G. Krespan*, J. Amer. Chem. Soc. *91*, 415 (1969).

[116] *H. Volz*, Tetrahedron Lett. *1963*, 3413; *1966*, 5229.

[117] *M. Ballester, J. Riera-Figueras*, and *A. Rodriguez-Siurana*, Tetrahedron Lett. *1970*, 3615.

[118] *R. West* and *P. T. Kwitowski*, J. Amer. Chem. Soc. *88*, 5280 (1966).

[119] *G. A. Olah* and *M. B. Comisarow*, J. Amer. Chem. Soc. *91*, 2955 (1969).

[120] *G. A. Olah*, *Y. K. Mo*, and *Y. Halpern*, J. Amer. Chem. Soc. *94*, 3551 (1972).

[121] *G. A. Olah* and *J. M. Bollinger*, J. Amer. Chem. Soc. *89*, 4744 (1967); *90*, 947 (1968); *G. A. Olah*, *J. M. Bollinger*, and *J. Brinich*, *ibid.* *90*, 2587 (1968).

[122] *G. A. Olah* and *P. E. Peterson*, J. Amer. Chem. Soc. *90*, 4675 (1968).

[123] *G. A. Olah* and *R. J. DeMember*, J. Amer. Chem. Soc. *91*, 2113 (1969); *92*, 718, 2562 (1970).

[124] For a review of the early literature see *P. Pfeiffer:* Organische Molekülverbindungen. F. Enke, Stuttgart 1927.

[125] *J. N. Collie* and *T. Tickle*, J. Chem. Soc. *75*, 710 (1899).

[126] *A. Hantzsch*, Chem. Ber. *55*, 953 (1922); *58*, 612, 941 (1925).

[127] *H. Meerwein*, *K. Bodenbenner*, *P. Borner*, *F. Kunert*, and *K. Wunderlich*, Liebigs Ann. Chem. *632*, 38 (1960); *H. Meerwein*, *V. Hederick*, *H. Morschel*, and *K. Wunderlich*, *ibid.* *635*, 1 (1960).

[128] *B. C. Ramsey* and *R. W. Taft*, *Jr.*, J. Amer. Chem. Soc. *88*, 3058 (1966).

[129] *G. A. Olah* and *J. M. Bollinger*, J. Amer. Chem. Soc. *89*, 2993 (1967).

[130] *C. MacLean* and *E. L. Mackor*, J. Chem. Phys. *34*, 2207 (1961).

[131] *T. Birchall* and *R. J. Gillespie*, Can. J. Chem. *43*, 1045 (1965).

[132] *H. Hogeveen*, Rec. Trav. Chim. Pays-Bas *86*, 696 (1967); *D. M. Brouwer*, *ibid.* *86*, 879 (1967).

[133] *M. Brookhart*, *G. C. Levy*, and *S. Winstein*, J. Amer. Chem. Soc. *89*, 1735 (1967).

[134] *G. A. Olah*, *D. H. O'Brien*, and *M. Calin*, J. Amer. Chem. Soc. *89*, 3582 (1967); *D. H. O'Brien*, *ibid.* *89*, 3586 (1967); *G. A. Olah* and *M. Calin*, *ibid.* *90*, 938 (1968).

[135] *G. A. Olah* and *A. M. White*, J. Amer. Chem. Soc. *91*, 5801 (1969).

[136] *G. A. Olah* and *A. M. White*, J. Amer. Chem. Soc. *89*, 3591, 4752 (1967); *H. Hogeveen*, Rec. Trav. Chim. Pays-Bas *86*, 289, 809 (1967).

[137] *G. A. Olah*, *D. H. O'Brien*, and *A. M. White*, J. Amer. Chem. Soc. *89*, 5694 (1967); *H. Hogeveen*, Rec. Trav. Chim. Pays-Bas *86*, 816 (1967); *H. Hogeveen*, *A. F. Bickel*, *C. W. Hilbers*, *E. L. Mackor*, and *C. MacLean*, Chem. Commun. *1966*, 898; Rec. Trav. Chim. Pays-Bas *86*, 687 (1967).

[138] *G. A. Olah* and *A. M. White*, J. Amer. Chem. Soc. *90*, 1884 (1968).

[139] *F. Seel*, Z. Anorg. Allg. Chem. *252*, 24 (1943).

[140] *D. Cook*, Can. J. Chem. *37*, 48 (1959); for a review see also *D. Cook* in *G. A. Olah:* Friedel-Crafts and Related Reactions. Interscience, New York 1963, Vol. 1, pp. 767ff.

[141] *B. P. Susz* and *J. J. Wuhrman*, Helv. Chim. Acta *40*, 971 (1957), and subsequent publications.

[142] *G. A. Olah, S. Kuhn*, and *S. Beke*, Chem. Ber. *89*, 862 (1956); *G. A. Olah, S. J. Kuhn, W. S. Tolgyesi*, and *E. B. Baker*, J. Amer. Chem. Soc. *84*, 2733 (1962); *G. A. Olah*, Rev. Chim. Acad. Rep. Populaire Roumaine 7, 1139 (1962); *W. S. Tolgyesi, S. J. Kuhn, M. E. Moffat, I. J. Bastien*, and *E. B. Baker*, J. Amer. Chem. Soc. *85*, 1328 (1963).

[143] *N. C. Deno, C. U. Pittman, Jr.*, and *M. J. Wisotsky*, J. Amer. *M. Calin, ibid. 90*, 938 (1968).

[144] *G. A. Olah* and *A. M. White*, J. Amer. Chem. Soc. *89*, 7072 (1967).

[145] *F. P. Boer*, J. Amer. Chem. Soc. *88*, 1572 (1966).

[146] *G. A. Olah* and *M. B. Comisarow*, J. Amer. Chem. Soc. *88*, 4442 (1966).

[147] *G. A. Olah* and *M. B. Comisarow*, J. Amer. Chem. Soc. *88*, 3313 (1966).

[148] *G. A. Olah* and *M. B. Comisarow*, J. Amer. Chem. Soc. *89*, 2694 (1967).

[149] *R. J. Gillespie* and *T. Birchall*, Can. J. Chem. *41*, 148 (1963).

[150] *V. C. Armstrong, D. W. Farlow*, and *R. B. Moodie*, Chem. Commun. *1968*, 1362.

[151] *G. A. Olah, A. M. White*, and *A. T. Ku*, J. Org. Chem. *36*, 3585 (1971).

[152] *G. A. Olah, D. L. Brydon*, and *R. D. Porter*, J. Org. Chem. *35*, 317 (1970).

[153] *J. L. Sudmeier* and *K. E. Schwartz*, Chem. Commun. *1968*, 1646.

[154] *G. A. Olah* and *A. M. White*, J. Amer. Chem. Soc. *90*, 6087 (1968).

[155] *G. A. Olah* and *T. E. Kiovsky*, J. Amer. Chem. Soc. *90*, 4666 (1968).

[156] *H. Meerwein, P. Laasch, R. Mersch*, and *J. Spille*, Chem. Ber. *89*, 209 (1956).

[157] *U. Müller* and *K. Dehnicke*, Angew. Chem. *78*, 825 (1966); Angew. Chem. internat. Edit. *5*, 841 (1966).

[158] *G. A. Olah, A. M. White*, and *D. H. O'Brien*, Chem. Rev. *70*, 561 (1970), and literature cited therein.

[159] *M. Saunders, E. L. Hagen*, and *J. Rosenfeld*, J. Amer. Chem. Soc. *90*, 6882 (1968).

[160] *M. Saunders, P. v. R. Schleyer*, and *G. A. Olah*, J. Amer. Chem. Soc. *86*, 5680 (1964).

[161] See lectures by *S. Winstein* and *G. A. Olah*, 21st Nat. Organic Chemistry Symposium ACS, Univ. of Utah, Salt Lake City, 1969, Abstract pp. 90, 143.

[162] a) *R. K. Lustgarten, M. Brookhart*, and *S. Winstein*, J. Amer. Chem. Soc. *89*, 6350 (1967); b) *M. Brookhart, A. Diaz*, and *S. Winstein*, *ibid.* *88*, 3135 (1966).

[163] *H. G. Richey, Jr.* and *R. K. Lustgarten*, J. Amer. Chem. Soc. *88*, 3136 (1966).

[164] a) *P. R. Story* and *M. Saunders*, J. Amer. Chem. Soc. *84*, 4876 (1962); b) *L. C. Snyder, D. C. Douglass, E. W. Anderson*, and *R. L. Kornegar*, *ibid.* *85*, 3630 (1963).

[165] *S. Winstein et al.*, J. Amer. Chem. Soc. *71*, 2953 (1949); *74*, 1154 (1952); *87*, 381 (1965), and intervening papers. Also for a summary see *S. Winstein*, Quart. Rev. Chem. Soc. *23*, 141 (1969).

[166] *J. D. Roberts* and *R. H. Mazur*, J. Amer. Chem. Soc. *73*, 3542 (1951); *R. H. Mazur, W. N. White, D. A. Semenow, C. C. Lee, M. S. Silver*, and *J. D. Roberts*, *ibid.* *81*, 4390 (1959).

[167] *K. B. Wiberg*, Tetrahedron *24*, 1083 (1968); *K. B. Wiberg* and *G. Szeimies*, J. Amer. Chem. Soc. *90*, 4195 (1968); for a review see *K. B. Wiberg, B. A. Hess, Jr.*, and *A. J. Ashe, III* in *G. A. Olah* and *P. v. R. Schleyer:* Carbonium Ions. Wiley-Interscience, New York, Vol. 3, in press.

[168] *D. J. Cram*, J. Amer. Chem. Soc. *71*, 3863 (1949) and subsequent publications; for a critical review see *D. J. Cram*, *ibid.* *86*, 3767 (1964).

[169] a) *G. A. Olah, M. B. Comisarow, E. Namanworth*, and *B. Ramsay*, J. Amer. Chem. Soc. *89*, 5259 (1967); b) *G. A. Olah, M. B. Comisarow*, and *C. J. Kim*, *ibid.* *91*, 1458 (1969); c) *G. A. Olah* and *R. D. Porter*, *ibid.* *92*, 7627 (1970); *93*, 6877 (1971).

[170] For comprehensive reviews see a) *C. K. Ingold:* Structure and Mechanism in Organic Chemistry. Cornell University Press Ithaca 1959, 2nd Edit.; b) *G. A. Olah:* Friedel-Crafts and Related Reactions. Wiley-Interscience, New York 1963—1965, Vol. 1—4; c) *D. Bethell* and *V. Gold:* Carbonium Ions, An Introduction. Academic Press, New York 1967; d) *G. A. Olah* and *P. v. R. Schleyer:* Carbonium Ions. Wiley-Interscience, New York 1968—1972, Vol. 1—3, Vol. 4 in press.

[171] a) *V. L. Talroze* and *A. L. Lyubimova*, Dokl. Akad. Nauk SSSR *86*, 909 (1952); b) *F. H. Field* and *M. S. B. Munson*, J. Amer. Chem. Soc. *87*, 3289 (1965), and literature cited therein.

[172] a) *G. A. Olah* and *J. Lukas*, J. Amer. Chem. Soc. *89*, 2227, 4739 (1967); *90*, 933 (1968); b) *G. A. Olah* and *R. H. Schlosberg*, *ibid. 90*, 2726 (1968); c) *H. Hogeveen* and *A. F. Bickel*, Chem. Commun. *1967*, 635; *H. Hogeveen*, *C. J. Gaasbeck*, and *A. F. Bickel*, Rec. Trav. Chim. Pays-Bas *88*, 703 (1969); *H. Hogeveen* and *C. J. Gaasbeek*, *ibid. 87*, 319 (1968).

[173] *G. A. Olah*, *G. Klopman*, and *R. H. Schlosberg*, J. Amer. Chem. Soc. *91*, 3261 (1969).

[174] a) *A. Gamba*, *G. Korosi*, and *M. Simonetta*, Chem. Phys. Lett. *3*, 20 (1969); b) *W. Th. van der Lugt* and *P. Ros*, *ibid. 4*, 389 (1969); c) *J. J. C. Mulder* and *J. S. Wright*, *ibid. 5*, 445 (1970); d) *H. Kollmar* and *H. O. Smith*, *ibid. 5*, 7 (1970).

[175] a) *V. Dyczmons*, *V. Staemmler* and *W. Kutzelnig*, Chem. Phys. Lett. *5*, 361 (1970); b) *A. Dedieu* and *A. Veillard*, Lecture at 21st Annual Meeting of the Société Chimique Physique, Paris, Sept. 1970; *A. Veillard*, personal communication; c) *W. A. Lathan*, *W. J. Hehre*, and *J. A. Pople*, Tetrahedron Lett. *1970*, 2699; d) J. Amer. Chem. Soc. *93*, 808 (1971).

[176] *E. L. Muetterties*, J. Amer. Chem. Soc. *91*, 1636 (1969).

[177] *R. E. Mesner* and *W. L. Jolly*, Inorg. Chem. *1*, 608 (1961).

[178] *G. A. Olah*, *P. W. Westerman*, *Y. K. Mo*, and *G. Klopman*, J. Amer. Chem. Soc. *94*, 7859 (1972).

[179] *A. N. Well*, *J. T. New*, and *K. S. Pitzer*, J. Chem. Phys. *17*, 1007 (1946).

[180] *G. A. Olah* and *J. A. Olah*, J. Amer. Chem. Soc. *93*, 1256 (1971).

[181] *G. A. Olah*, *Y. Halpern*, *J. Shen*, and *Y. K. Mo*, J. Amer. Chem. Soc. *93*, 1251 (1971).

[182] *G. A. Olah* and *H. C. Lin*, J. Amer. Chem. Soc. *93*, 1259 (1971).

[183] *G. A. Olah*, *R. Renner*, and *P. Schilling*, to be published.

[184] *G. A. Olah* and *G. D. Mateescu*, 159th Nat. Meet. Amer. Chem. Soc., Houston, Texas, Feb. 1970, Abstr. Papers.

[185] a) *E. S. Lewis* and *M. C. R. Symmons*, Quart. Rev. Chem. Soc. *12*, 230 (1958); b) *M. F. Hawthorne* and *E. S. Lewis*, J. Amer. Chem. Soc. *80*, 4296 (1958).

[186] *G. A. Olah*, *J. Shen*, and *R. H. Schlosberg*, J. Amer. Chem. Soc. *92*, 3831 (1970).

[187] a) *H. Hogeveen, C. J. Gaasbeek,* and *A. F. Bickel,* Rev. Trav. Chim. Pays-Bas *88,* 716 (1969); b) *D. M. Brouwer* and *H. Hogeveen,* Progr. Phys. Org. Chem. *2,* 179 (1972).

[188] *A. F. Bickel, C. J. Gaasbeek, H. Hogeveen, J. N. Oelderik,* and *J. C. Platteeuw,* Chem. Commun. *1967,* 634.

[189] *L. Schmerling,* J. Amer. Chem. Soc. *66,* 1422 (1944); *67,* 1778 (1945); *68,* 153 (1946).

[190] a) *P. D. Bartlett, F. E. Condon,* and *A. Schneider,* J. Amer. Chem. Soc. *66,* 1531 (1944); b) *C. D. Nenitzescu, M. Avram,* and *E. Sliam,* Bull. Soc. Chim. France *1955,* 1266, and literature cited therein.

[191] *G. A. Olah, J. R. DeMember,* and *R. H. Schlosberg,* J. Amer. Chem. Soc. *91,* 2112 (1969).

[192] *G. A. Olah, J. R. DeMember, R. H. Schlosberg,* and *Y. Halpern,* J. Amer. Chem. Soc. *94,* 156 (1972).

[193] a) *G. A. Olah* and *J. R. DeMember,* J. Amer. Chem. Soc. *91,* 2113 (1969); b) *92,* 718 (1970); c) *92,* 2562 (1970).

[194] An electron deficient atom is defined as one which has more bonding orbitals than electrons in its valence shell (*R. E. Rundle,* J. Amer. Chem. Soc. *69,* 1327 (1947); J. Chem. Phys. *17,* 671 (1949)). In a carbonium ion two electrons from the original $\sigma$-orbital are delocalized over three centers.

[195] For a review of isomerization of alkanes see *H. Pines* and *N. E. Hoffmann* in *G. A. Olah:* Friedel-Crafts and Related Reactions, Wiley-Interscience, New York 1964, Vol. 2, Chap. 28.

[196] a) *C. D. Nenitzescu* and *I. P. Cantuniari,* Chem. Ber. *66,* 1097 (1933); b) *C. D. Nenitzescu, M. Avram,* and *E. Sliam,* Bull. Soc. Chim. France *1955,* 1266.

[197] *D. P. Stevenson* and *J. H. Morgan,* J. Amer. Chem. Soc. *70,* 2890 (1948).

[198] *G. A. Olah* and *J. Lukas,* J. Amer. Chem. Soc. *90,* 933 (1968).

[199] *F. C. Whitmore,* J. Amer. Chem. Soc. *54,* 3274 (1932); *F. C. Whitmore* and *E. E. Stahly, ibid. 55,* 4153 (1933); *F. C. Whitmore,* Chem. Eng. News *26,* p. 668 (March 8, 1948).

[200] *P. v. R. Schleyer et al.,* J. Amer. Chem. Soc. *92,* 5246 (1970).

[201] *D. M. Brouwer* and *H. Hogeveen,* Rec. Trav. Chim. Pays-Bas *89,* 211 (1970).

[202] a) *N. C. Deno, G. Saines,* and *M. Spangler,* J. Amer. Chem. Soc. *84,* 3295 (1962); b) *R. Stewart:* Oxidation Mechanisms. Benjamin, New York 1966, p. 22.

[203] *H. J. Dauben, Jr., F. A. Gadecki, K. M. Harmon,* and *D. L. Pearsen*, J. Amer. Chem. Soc. *79*, 4557 (1957).

[204] *H. J. Dauben, Jr.* and *D. J. Beitelli*, J. Amer. Chem. Soc. *83*, 4657, 4659 (1961); for a summary see *K. M. Harmon, G. A. Olah,* and *P. v. R. Schleyer:* Carbonium Ions. Wiley-Interscience, New York 1972, Vol. 3, and literature cited therein.

[205] Experimental results supporting this suggestion were recently reported by *G. A. Olah* and *J. J. Svoboda*, Manchester Meeting of the Chemical Society, April 1972; J. Amer. Chem. Soc., in press.

[206] *H. Lankamp, W. T. Nauta,* and *C. MacLean*, Tetrahedron Lett. *1968*, 249.

[207] For general discussion see, for example, *J. March:* Advanced Organic Chemistry Reactions, Mechanisms and Structure. McGraw-Hill, New York 1968.

[208] *R. W. Taft, Jr.* (J. Amer. Chem. Soc. *74*, 5372 (1952)) seems to be the first to have suggested the initial formation, in the acid catalyzed hydration of isobutylene, of a proton π-complex of the type described originally by *M. J. S. Dewar* (J. Chem. Soc. *1946*, 406 and Bull. Soc. Chim. France *1951*, C 75) in which "...the proton is embedded in the π-orbitals which extend above (or below) the plane of the C—C double bond".

[209] For theoretical aspects see *L. Radom, J. A. Pople, V. Buss,* and *P. v. R. Schleyer*, J. Amer. Chem. Soc., in press.

[209a] *T. G. Traylor, W. Hanstein, H. J. Berwin, N. A. Clinton,* and *R. S. Brown*, J. Amer. Chem. Soc. *93*, 5715 (1971); *W. Hanstein, H. J. Berwin,* and *T. G. Traylor*, *ibid. 92*, 829 (1970).

[210] *G. A. Olah, J. M. Bollinger,* and *J. Brinich*, J. Amer. Chem. Soc. *90*, 2587 (1968); *G. A. Olah* and *J. M. Bollinger*, *ibid. 89*, 4744 (1967); *90*, 947 (1968).

[211] *G. A. Olah* and *P. R. Clifford*, J. Amer. Chem. Soc. *93*, 1261, 2320 (1971).

[212] *J. Strating, J. H. Wieringa,* and *H. Wynberg*, Chem. Commun. *1969*, 907.

[212a] *G. A. Olah et al.*, to be published.

[213] *H. C. Brown:* Hydroboration. Benjamin, New York 1962.

[214] *H. Lehmkuhl* and *K. Ziegler* in *Houben-Weyl-Müller:* Methoden der Organischen Chemie. Thieme, Stuttgart 1970, 4th Edit., Vol. 13/4, pp. 1ff.

[215] *G. S. Hammond*, J. Amer. Chem. Soc. 77, 334 (1955).

[216] *G. A. Olah*, Accounts Chem. Res. *4*, 240 (1971).

[217] *G. A. Olah, H. Lin*, and *Y. K. Mo*, to be published.

[218] a) *M. J. S. Dewar*, J. Chem. Soc. *1946*, 406; b) Bull. Soc. Chim. Francc *1951*, C 75; c) Electronic Theory of Organic Chemistry. Oxford University Press 1949.

[219] *M. J. S. Dewar* and *A. P. Marchand*, Annu. Rev. Phys. Chem. *16*, 321 (1965).

[220] For a review of *Ingold*'s publications see *C. K. Ingold:* Structure and Mechanism in Organic Chemistry. 2nd Edit., Cornell University Press, Ithaca, N.Y., 1969.

[221] *S. Winstein* and *H. J. Lucas*, J. Amer. Chem. Soc. *61*, 1576 (1939); see also *I. Roberts* and *G. E. Kimball*, *ibid. 59*, 947 (1937).

[222] *P. D. Bartlett:* Nonclassical Ions. Reprints and Commentary. Benjamin, New York 1965, and the numerous references cited therein.

[223] *D. J. Cram*, J. Amer. Chem. Soc. *71*, 3863 (1949) and subsequent publications; see also *D. J. Cram*, *ibid. 86*, 3767 (1964).

[224] *M. Stiles* and *R. P. Mayer*, J. Amer. Chem. Soc. *81*, 1497 (1959).

[225] *J. Berson* and *S. Suzuki*, J. Amer. Chem. Soc. *81*, 4088 (1959).

[226] *J. E. Williams, Jr., V. Buss, L. C. Allen, P. v. R. Schleyer, W. A. Lathan, W. J. Hehre*, and *J. A. Pople*, J. Amer. Chem. Soc. *92*, 2141 (1970); *G. V. Pfeiffer* and *J. C. Jewett*, *ibid. 92*, 2143 (1970).

[227] *G. N. Lewis*, J. Amer. Chem. Soc. *38*, 762 (1916); Valence and Structure of Atoms and Molecules. Chemical Catalog Corp., New York 1923.